# Organic Fluorine Chemistry

# Organic Fluorine Chemistry

**Miloš Hudlický**

*Department of Chemistry*
*Virginia Polytechnic Institute and State University*
*Blacksburg, Virginia*

PLENUM PRESS • NEW YORK–LONDON • 1971

Library of Congress Catalog Card Number 76-131889
SBN 306-30488-0


A Division of Plenum Publishing Corporation
227 West 17th Street, New York, N.Y. 10011

United Kingdom edition published by Plenum Press, London
A Division of Plenum Publishing Company, Ltd.
Davis House (4th Floor), 8 Scrubs Lane, Harlesden, NW10 6SE, England



Printed in the United States of America

**To My Mother**

# Preface

The present book is essentially based on the lectures on the chemistry of organic compounds of fluorine that I gave in 1969 at Virginia Polytechnic Institute in Blacksburg, Virginia, as a graduate course. References to material published to the end of 1969 are included. The book is primarily meant to provide the background for such a course, and, at the same time, to be a brief survey of recent knowledge in, and an introduction to deeper study of, this area of chemistry, which has been treated in a number of comprehensive monographs.

I would like to thank Professor S. C. Cohen, Syracuse University, for the compilation of the data on mass spectra and nuclear magnetic resonance spectra, and my son, Tomáš Hudlický, and my daughter, Eva Hudlická, for their help with the indexes.

Miloš Hudlický
*Virginia Polytechnic Institute and State University Blacksburg, Virginia*

February 13, 1970

# Contents


CHAPTER 1. *Introduction* ........ 1
Development of Fluorine Chemistry ........ 1
Handling of Fluorine, Hydrogen Fluoride, and Fluorine Compounds ........ 5
Equipment and Apparatus ........ 6

CHAPTER 2. *Fluorinating Agents* ........ 9
Hydrogen Fluoride ........ 12
Fluorine ........ 13
Inorganic Fluorides ........ 14
Organic Fluorides ........ 14

CHAPTER 3. *Nomenclature of Organic Fluorine Compounds* ........ 18

CHAPTER 4. *Introduction of Fluorine into Organic Compounds* ........ 21
Addition of Hydrogen Fluoride to Olefins ........ 21
Addition of Hydrogen Fluoride to Acetylenes and Other Unsaturated Systems ........ 22
Addition of Fluorine to Olefins ........ 23
Addition of Fluorine to Other Unsaturated Systems ........ 24
Addition of Halogen Fluorides and Organic Fluorides to Olefins ........ 25
Replacement of Hydrogen by Fluorine ........ 28
Replacement of Halogens by Fluoride ........ 31
Replacement by Means of Hydrogen Fluoride ........ 33
Replacement by Means of Antimony Fluorides ........ 33
Replacement by Means of Silver Fluoride ........ 35
Replacement by Means of Mercury Fluorides ........ 35
Replacement by Means of Potassium Fluoride ........ 36
Replacement by Means of Other Fluorides ........ 37
Replacement of Oxygen by Fluorine ........ 38
Cleavage of Ethers and Epoxides ........ 38
Cleavage of Sulfonic Esters ........ 39

Replacement of Hydroxylic Group by Fluorine ........ 39
Replacement of Carbonyl Oxygen by Fluorine ........ 41
Conversion of Carboxylic Group to Trifluoromethyl Group ........ 41
Replacement of Nitrogen by Fluorine ........ 42

CHAPTER 5. *Analysis of Organic Fluorine Compounds* ........ 46
Physical Methods of Analysis of Fluorine Compounds or Their Mixtures ........ 46
Infrared Spectroscopy ........ 46
Mass Spectroscopy ........ 47
Nuclear Magnetic Resonance Spectroscopy ........ 48
Separation of Mixtures and Purification of Components ........ 50
Qualitative Tests for Fluorine ........ 50
Detection of Elemental Fluorine ........ 50
Detection of Fluoride Ion ........ 51
Detection of Fluorine in Organic Compounds ........ 51
Quantitative Determination of Fluoride ........ 52
Volumetric Determination ........ 52
Determination Using Fluoride Membrane Electrode ........ 53
Mineralization of Fluorine in Organic Compounds ........ 53
Decomposition with Alkalies ........ 53
Combustion in Oxygen ........ 54
Determination of Other Elements in Organic Fluorine Compounds ........ 54
Special Analyses ........ 55

CHAPTER 6. *Properties of Organic Fluorine Compounds* ........ 56
Physical Properties ........ 57
Melting Points ........ 57
Boiling Points ........ 57
Density ........ 60
Refractive Index ........ 60
Dielectric Constant ........ 61
Surface Tension ........ 61
Viscosity ........ 63
Solubility ........ 63
Physicochemical Properties ........ 63
Biological Properties ........ 66

CHAPTER 7. *Practical Applications of Organic Fluorine Compounds* ........ 69

Refrigerants, Propellants, and Fire Extinguishers .......... 69
Plastics and Elastomers .......... 70
Monomers .......... 72
Polymerization .......... 74
Processing of Polymers .......... 74
Applications of Plastics and Elastomers .......... 74
Fluorinated Compounds as Pharmaceuticals .......... 77
Other Uses of Fluorinated Compounds .......... 78

CHAPTER 8. *Reactions of Organic Fluorine Compounds* .......... 80
Factors Governing the Reactivity of Organic Fluorine Compounds .......... 80
Inductive Effect .......... 80
Hyperconjugation .......... 84
Mesomeric Effect .......... 84
Steric Effects .......... 85
Important Features in the Reactivity of Organic Fluorine Compounds .......... 86
Reduction .......... 88
Catalytic Hydrogenation .......... 89
Reduction with Complex Hydrides .......... 90
Reduction with Metals and Metallic Compounds .......... 93
Reduction with Organic Compounds .......... 93
Oxidation .......... 95
Oxidations with Oxygen .......... 95
Oxidations with Oxidative Reagents .......... 95
Oxidations of Fluoro-olefins .......... 95
Oxidations of Fluorinated Aromatics .......... 96
Oxidation of Nitrogen and Sulfur Compounds .......... 97
Anodic Oxidation .......... 98
Electrophilic Reactions .......... 98
Halogenation .......... 98
Addition of Halogens across Multiple Bonds .......... 99
Replacement of Hydrogen by Halogens .......... 99
Cleavage of the Carbon Chain by Halogens .......... 101
Addition of Hydrogen Halides across Multiple Bonds .......... 101
Replacement of Oxygen- and Nitrogen-Containing Functions by Halogens .......... 102
Replacement of Fluorine by Other Halogens .......... 102
Nitration .......... 103
Nitrosation .......... 104
Sulfonation .......... 104

Acid-Catalyzed Syntheses (Friedel-Crafts Reaction) .................. 104
Acid-Catalyzed Additions .................. 105
Acid-Catalyzed Substitutions .................. 105
Nucleophilic Substitutions .................. 106
Esterification and Acetalization .................. 106
Hydrolysis .................. 108
Hydrolysis of Nonfluorinated Parts of Fluorinated Molecules .................. 108
Hydrolytic Displacement of Single Fluorine Atoms .................. 108
Hydrolysis of Difluoromethylene Group .................. 109
Hydrolysis of Trifluoromethyl Group .................. 110
Hydrolysis of Perfluoro Compounds .................. 110
Fluoroform Reaction .................. 111
Alkylations .................. 111
Alkylations at Oxygen .................. 111
Alkylations at Sulfur .................. 113
Alkylations at Nitrogen and Phosphorus .................. 114
Alkylations at Carbon .................. 115
Arylations .................. 115
Arylations at Oxygen .................. 115
Arylations at Sulfur .................. 117
Arylations at Nitrogen .................. 117
Arylations at Carbon .................. 118
Acylations .................. 118
Acylations at Oxygen .................. 118
Acylations at Sulfur .................. 119
Acylations at Nitrogen .................. 119
Acylations at Carbon .................. 119
Syntheses with Organometallic Compounds .................. 120
The Grignard Syntheses .................. 120
Grignard Reagents as Organic Substrate .................. 121
Fluorinated Grignard Reagents .................. 121
Perfluorinated Grignard Reagents .................. 122
Organolithium Compounds .................. 125
Organolithium Compounds as Organic Substrate .................. 125
Fluorinated Organolithium Compounds .................. 126
Perfluorinated Organolithium Compounds .................. 127
Organozinc Compounds .................. 128
Organomercury Compounds .................. 128
Organocopper Compounds .................. 129
Organometalloids .................. 129
Base-Catalyzed Condensations .................. 129

Additions .......... 133
Nucleophilic Additions to Fluorinated Olefins .......... 133
Additions of Alcohols and Phenols .......... 133
Additions of Mercaptans and Thiophenols .......... 133
Additions of Ammonia and Amines .......... 134
Free-Radical-Type Additions .......... 134
Linear Additions .......... 134
Cycloadditions .......... 138
Formation of Three-Membered Rings .......... 138
Formation of Four-Membered Rings .......... 138
Formation of Five-Membered Rings .......... 139
Formation of Six-Membered Rings (Diels–Alder Reaction) .......... 140
Eliminations .......... 142
Dehalogenations .......... 142
Dehydrohalogenations .......... 143
Decarboxylations .......... 144
Dehydration .......... 145
Molecular Rearrangements .......... 145
Pyroreactions .......... 147

CHAPTER 9. *Fluorinated Compounds as Chemical Reagents* .......... 152

*References* .......... 154

*Author Index* .......... 171

*Subject Index* .......... 181

*Chapter 1*

# Introduction

## DEVELOPMENT OF FLUORINE CHEMISTRY

The chemistry of fluorine has been marked with a certain delay in development as compared with the chemistry of other halogens. Hydrofluoric acid was not prepared until 1771, by K. W. Scheele, and elemental fluorine not until 1886, when H. Moissan subjected a solution of potassium fluoride in anhydrous hydrogen fluoride to electrolysis in a platinum apparatus. This delay is reflected in the chemistry of fluorine compounds. The early development of inorganic fluorine chemistry is mainly due to H. Moissan and O. Ruff. The foundations of organic fluorine chemistry were laid by F. Swarts in the late 19th and early 20th century. It was not, however, until 1930 that Freons, fluorinated refrigerants, were discovered by T. Midgley, Jr., and A. L. Henne, and thus organic fluorine compounds entered the field of large-scale technology. This important discovery started a kind of a chain reaction which led to the development of the technology of elemental fluorine, of many methods of preparation of inorganic and especially organic fluorine compounds, and to the discovery of important and peculiar compounds such as Teflon and other fluorocarbons. Fluorine chemistry played an important role in atomic energy projects (isotopic uranium hexafluorides were separated by thermodiffusion in the medium of fluorocarbons), and plays a part even in space research.

Fluorine chemistry caused an "explosion" not only in the number of prepared compounds, which increased over a period of 30 years (1930–1960) from several hundreds to over 14,000, but also in the number of publications on this subject, which is ever increasing. There were three monographs on fluorine prior to 1930, whereas there are more than 30 today.

In order to facilitate the exchange of views and information, international symposia have been established: 1959, Birmingham, England; 1962, Estes Park, USA; 1965, Munich, Germany; 1967, Estes Park, USA; 1969, Moscow, USSR; 1971, Durham, England.

Table 1 lists the most essential monographs now available to the fluorine

**Table 1. Selected Monographs on Fluorine Chemistry**

| Author or Editor | Title | Publisher | Place | Year |
|---|---|---|---|---|
| Haszeldine, R. N., and Sharpe, A. G. | Fluorine and Its Compounds | Methuen and Co. | London | 1951 |
| Lovelace, A. M., Rausch, D. A., and Postelnek, W. | Aliphatic Fluorine Compounds | Reinhold Pub. Corp. | New York | 1958 |
| Pavlath, A. E., and Leffler, A. L. | Aromatic Fluorine Compounds | Reinhold Pub. Corp. | New York | 1962 |
| Forche, E., Hahn, W., and Stroh, R. | Fluorverbindungen: Herstellung, Reaktivität und Unwandlungen (Houben-Weyl, Methoden der Organischen Chemie, 5/3) | G. Thieme Verlag | Stuttgart | 1962 |
| Sheppard, W. A., and Sharts, C. M. | Organic Fluorine Chemistry | W. A. Benjamin | New York | 1969 |
| Simons, J. H. | Fluorine Chemistry, I–V | Academic Press | New York | 1950<br>1954<br>1963<br>1965<br>1964 |
| Stacey, M., Tatlow, J. C., and Sharpe, A. G. | Advances in Fluorine Chemistry, 1–5 | Butterworths | London | 1960<br>1961<br>1963<br>1965<br>1965 |
| Tarrant, P. | Fluorine Chemistry Reviews, 1–4 | Marcel Dekker | New York | 1967<br>1968<br>1969 |
| Pattison, F. L. M. | Toxic Aliphatic Fluorine Compounds | Elsevier | Amsterdam | 1959 |
| Banks, R. E. | Fluorocarbons and Their Derivatives | Oldbourne Press | London | 1963 |

**Table 2. Topics Covered in the Main Series on Fluorine Chemistry**

**Fluorine Chemistry,** J. H. Simons, editor

*Volume I*

Nonvolatile Inorganic Fluorides. Emeleus, H. J.
Volatile Inorganic Fluorides. Burg, A. B.
The Chemistry of the Fluoro Acids of Fourth, Fifth, and Sixth Group Elements. Lange, W.
The Halogen Fluorides. Booth, H. S.
Boron Trifluoride. Booth, H. S.
Hydrogen Fluoride. Simons, J. H.
Hydrogen Fluoride Catalysis. Simons, J. H.
Preparation of Fluorine. Cady, G. H.
Physical Properties of Fluorine. Cady, G. H.
The Theoretical Aspects of Fluorine Chemistry. Glockler, G.
The Action of Elementary Fluorine upon Organic Compounds. Bigelow, L. A.
Fluorocarbons and Their Production. Simons, J. H.
Fluorocarbons–Their Properties and Wartime Development. Brice, T. J.
Fluorocarbon Derivatives. Pearlson, W. H.
Aliphatic Chlorofluoro Compounds. Park, J. D.
Fluorine Compounds in Glass Technology and Ceramics. Weyl, W. A.

*Volume II*

Fluorine-Containing Complex Salts and Acids. Sharpe, A. G.
Halogen Fluorides—Recent Advances. Emeleus, H. J.
Analytical Chemistry of Fluorine and Fluorine-Containing Compounds. Elving, P. J.
Organic Compounds Containing Fluorine. Tarrant, P.
Metallic Compounds Containing Fluorocarbon Radicals and Organometallic Compounds Containing Fluorine. Emeleus, H. J.
Fluorocarbon Chemistry. Simons, J. H.
The Infrared Spectra of Fluorocarbons and Related Compounds. Weiblen, D. G.

*Volume III*

Biological Effects of Fluorine Compounds. Hodge, H. C., Smith, F. A., and Chen, P. S.

*Volume IV*

Biological Properties of Inorganic Fluorides. Hodge, H. C., and Smith, F. A.
Effect of Fluorides on Bones and Teeth. Hodge, H. C., and Smith, F. A.

*Volume V*

General Chemistry of Fluorine-Containing Compounds. Simons, J. H.
Physical Chemistry of Fluorocarbons. Reed, T. M., III
Radiochemistry and Radiation Chemistry of Fluorine. Wethington, J. A., Jr.
Industrial and Utilitarian Aspects of Fluorine Chemistry. Brice, H. G.

**Advances in Fluorine Chemistry,** Stacey, M., Tatlow, J. C., and Sharpe, A. B., editors

*Volume I*

The Halogen Fluorides—Their Preparation and Uses in Organic Chemistry. Musgrave, W. K. R.
Transition-Metal Fluorides and Their Complexes. Sharpe, A. G.
The Electrochemical Process for the Synthesis of Fluoro-Organic Compounds. Burdon, J.

**Table 2 (Continued)**

Fluoroboric Acids and Their Derivatives. Sharp, D. W. A.
Exhaustive Fluorinations of Organic Compounds with High-Valency Metallic Fluorides. Stacey, M., and Tatlow, J. C.

*Volume 2*

The Thermochemistry of Organic Fluorine Compounds. Patrick, C. R.
Fluorine Resources and Fluorine Utilization. Finger, G. C.
Mass Spectrometry of Fluorine Compounds. Majer, J. R.
The Fluorides of the Actinide Elements. Hodge, N.
The Physiological Action of Organic Compounds Containing Fluorine. Saunders, B. C.
The Fluorination of Organic Compounds Using Elementary Fluorine. Tedder, J. M.

*Volume 3*

Effects of Adjacent Perfluoroalkyl Groups on Carbonyl Reactivity. Braendlin, H. P., and McBee, E. T.
Perfluoroalkyl Derivatives of the Elements. Clark, H. C.
Mechanisms of Fluorine Displacement. Parker, R. E.
Nitrogen Fluorides and Their Inorganic Derivatives. Colburn, C. B.
The Organic Fluorochemicals Industry. Hamilton, J. M., Jr.
The Preparation of Organic Fluorine Compounds by Halogen Exchange. Barbour, A. K.

*Volume 4*

The Balz–Schiemann Reaction. Suschitzky, H.
Some Techniques and Methods of Inorganic Fluorine Chemistry. Peacock, R. D.
Ionic Reactions of Fluoro-Olefins. Chambers, R. D.
Structural Aspects of Monofluorosteroids. Taylor, N. F., and Kent, P. W.
Fluorides of the Main Group Elements. Kemmitt, R. D. W., and Sharp, D. W. A.
The Vibrational Spectra of Organic Fluorine Compounds. Brown, J. K., and Morgan, K. J.

*Volume 5*

Oxyfluorides of Nitrogen. Woolf, C.
Fluorides of Phosphorus. Schmutzler, R.

**Fluorine Chemistry Reviews,** Tarrant, P., Richardson, R. D., and Lagowski, J. J., editors

*Volume 1*

Synthesis, Compounding, and Properties of Nitroso Rubbers. Henry, M. C., Griffis, C. B., and Stump, E. C.
Electrochemical Fluorination. Nagase, Shunhi
The Fluoroketenes. Cheburkov, Y. A., and Knunyants, I. L.
Hexafluoroacetone. Krespan, C. G., and Middleton, W. J.
Fluorocarbon Toxicity and Biological Action. Clayton, J. W.
Diels–Alder Reactions of Organic Fluorine Compounds. Perry, D. R. A.
Methods of the Introduction of Hydrogen into Fluorinated Compounds. Mettille, F. J., and Burton, D. J.
Reactions Involving Fluoride Ion and Polyfluoroalkyl Anions. Young, J. A.

*Volume 2*

The Cycloaddtion Reactions of Fluoroolefins. Sharkey, W. H.

Table 2 (Continued)

The Reactions of Halogenated Cycloalkenes with Nucleophiles. Park, J. D., Murtry, R. J., and Adams, J. H.
Ionization Potentials and Molecule–Ion Dissociation Energies for Diatomic Metal Halides. Hastie, J. W., and Margrave, J. L.
Nuclear Magnetic Resonance Spectra of M–F Compounds. Brey, W. L., and Hynes, J. L.
The $F^{19}$ Chemical Shifts and Coupling Constants of Fluoroxy Compounds. Hoffman, C. J.

*Volume 3*
Fluorine Compounds in Anesthesiology. Larsen, E. R.
Reactions of Fluoroolefins with Electrophilic Reagents. Dyatkin, B. L., Mochalina, E. P., and Knunyants, I. L.
Fluoroalicyclic Deirvatives of Metals and Metalloids. Cullen, W. R.
Phosphorus, Arsenic, and Antimony Pentafluorophenyl Compounds. Fild, M., and Glemser, O.

*Volume 4*
Polyhaloalkyl Derivatives of Sulfur. Dresdner, R. D., and Hooper, T. R.
The Chemistry of Fluorinated Acetylenes. Bruce, M. I., and Cullen, W. R.
The Chemistry of Aliphatic Fluoronitrocarbons. Bissell, E. R.

chemist, and Table 2 titles of the chapters in the collections *Fluorine Chemistry*, Volumes I–V, *Advances in Fluorine Chemistry*, Volumes 1–5, and *Fluorine Chemistry Reviews*, Volumes 1–4.

## HANDLING OF FLUORINE, HYDROGEN FLUORIDE, AND FLUORINE COMPOUNDS

Fluorine, hydrogen fluoride, and some inorganic and organic fluorides are poisonous, highly corrosive, and generally dangerous, and require special precautions in work with them.

One of the most hazardous substances is *elemental fluorine*, which may cause explosions in contact with organic material. It is imperative to use eye protection (plastic shield), rubber gloves, and a plastic apron when working with fluorine. Breathing even small concentrations of fluorine should be avoided.

Another dangerous compound requiring the same precautions is *anhydrous hydrogen fluoride*, especially as a liquid. Its action on skin is immediate and causes painful and slowly healing wounds. Even aqueous hydrofluoric (and also fluoroboric) acid injures skin readily, and any contact with fingers and nails should be carefully avoided.

Burns caused by fluorine or hydrogen fluoride must be immediately taken care of. The burnt area must be washed free of acid by tap water, if possible, immersed in ice-cold 70% ethyl alcohol for up to 30 min, and finally covered by a paste made from magnesium oxide and glycerol. With large and deep burns, subcutaneous injections of calcium gluconate in the injured area are recommended. When hydrogen fluoride gets into the eye, the eye is to be flushed with lukewarm water and finally with 2% solution of sodium bicarbonate. With serious injuries of the eye, 0.5% pantocain solution is to be applied to relieve the pain **[1]**.

Other fluorine compounds are much less dangerous, although handling them requires precautions usual for common poisonous and corrosive substances. This applies to most of the inorganic fluorides, especially to arsenic and antimony fluorides, and to strong acids such as fluoro-, difluoro-, and trifluoroacetic acid, fluoroboric acid, etc.

*Fluoroacetic acid* is dangerous not only by token of its acidity but also as an enzymic poison of high toxicity. For the same reason, its derivatives and homologs having an even number of carbon atoms in their chains and a single fluorine atom in the $\omega$-positions must also be handled with proper care **[2]**.

Another group of compounds, *alkyl fluorophosphates*, known as "nerve gases," are very poisonous because of their action on the enzyme cholinesterase **[3]**.

Although some of the fluorinated organic compounds such as *perfluoroisobutylene* are extremely toxic (of the order of phosgene), there is usually little chance of their being encountered. The only common source of perfluoroisobutylene is pyrolysis of polytetrafluoroethylene (Teflon), but this is serious only at very high temperatures, usually not encountered under the operating conditions of Teflon-coated tools **[4]**.

## EQUIPMENT AND APPARATUS

Common *glass equipment* can be used for working with most fluorides and even fluorine provided no hydrogen fluoride is generated during the reaction. Hydrogen fluoride, especially highly concentrated or anhydrous, eats up glass very rapidly, and its contact with glass must be limited to a very short time. Since silicon tetrafluoride is formed from glass and hydrogen fluoride, glass cannot be used in cases where such contamination is undesirable. For safety reasons, work with elemental fluorine or highly corrosive fluorides such as halogen fluorides and antimony pentafluoride is preferably done in plastics or metals. Thus, the use of glass apparatus is practically limited to work with alkaline fluorides, silver fluoride, mercury fluorides, antimony trifluoride, and a few others.

For work with all fluorides, including anhydrous hydrogen fluoride, at atmospheric or not very elevated pressure, *plastics* are very suitable provided not too high a temperature is used. The temperature limits for plastics are

**Table 3. Corrosion of Materials by Fluorine and Hydrogen Fluoride (Penetration in inches per year) [5]**

| Material | Dry fluorine | Anhydrous hydrogen fluoride | | Hydrofluoric acid 40–65%, |
|---|---|---|---|---|
| | | up to 50°C | 100–150°C | up to 50°C |
| Fe | <0.004 | <0.02 | >0.05 | >0.05 |
| Hastelloy | — | <0.02 | <0.02 | <0.02 |
| (55% Ni, 17% Mo, 16% Cr) | | | | |
| 6% Fe, 4% W | | | | |
| Ni | <0.004 | <0.02 | — | 0.02 |
| Monel | 0.004 | 0.02 | <0.02 | <0.02 |
| (67% Ni, 30% Cu, 3% Al) | | | | |
| Ag, Au, Pt | <0.004 | — | — | — |
| Cu | <0.004 | <0.02 | <0.05 | >0.05 |
| Pb | <0.04 | >0.05 | >0.05 | <0.004 |
| Bronze | — | <0.02 | >0.05 | >0.05 |
| Al | >0.1 | — | — | >0.05 |
| Glass | <0.004 | — | — | >0.05 |
| Plastics: | | | | |
| Epoxy resins | — | — | — | Stable |
| Polyester, polyamide | — | Unstable | — | Unstable |
| Polyvinyl chloride | — | Limited use | — | — |
| Polyethylene | Stable | Stable | — | Stable |
| Polychlorotrifluoroethylene | Stable | Stable | — | Stable |
| Polytetrafluoroethylene | Stable | Stable | — | Stable |

up to 100°C for polyethylene, about 200°C for polychlorotrifluoroethylene, and about 250–280°C for polytetrafluoroethylene or copolymers of tetrafluoroethylene with hexafluoropropylene.

Any kind of reaction of fluorides and fluorine, including reactions much above atmospheric pressure, can be carried out in *metallic equipment or autoclaves*. For most purposes, mild steel is satisfactory, although its corrosion may sometimes by very high, particularly in the presence of water in the reaction medium. Special stainless steel such as Hastelloy is much more resistant and more suitable, especially for work of high accuracy. Hydrogen fluoride, fluorine, and corrosive fluorides can be successfully handled in copper, and some fluorides even in aluminum equipment. The most suitable metals for work with hydrogen fluoride and fluorine are nickel and Monel metal, whose corrosion is very low even at high temperatures.

Numerical data in Table 3 show corrosion of various materials and can be a guide for the choice of material for equipment in any particular case [5].

In addition to apparatus fitted for any particular reaction, some special apparatus was developed for the preparation of organic fluorine compounds. Noncatalytic fluorinations with elemental fluorine require a special type of a

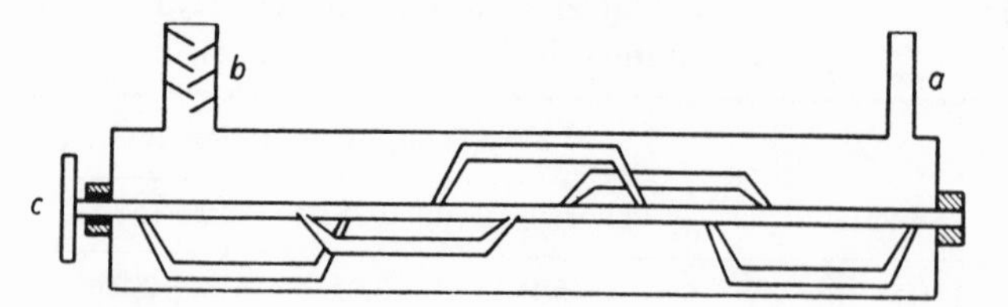


**Fig. 1.** Rotary apparatus for fluorination by means of silver difluoride or cobalt trifluoride: (a) inlet tube, (b) outlet tube with baffle plates, (c) rotating shaft with paddles.

cold-flame burner allowing fast mixing of the reaction components **[6]**. Catalytic fluorinations with elemental fluorine are carried out in vertical reactors filled with the appropriate substrate coated with the metal catalyst **[7]**. Reactions of organic compounds with high-valency metal fluorides such as cobalt trifluoride are carried out in rotary horizontal tubes fitted with paddle stirrers (Fig. 1) **[8]**. Electrochemical fluorinations achieved by electrolysis of organic compounds in liquid anhydrous hydrogen fluoride are performed in electrolytic cells similar to those used for the production of fluorine (Fig. 2) **[9]**.

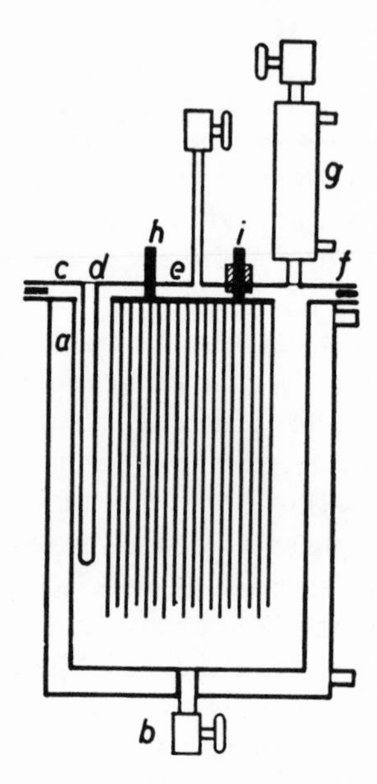


**Fig. 2.** Apparatus for electrolytic fluorination: (a) electrolytic cell with water jacket, (b) liquid product outlet, (c) cell lid, (d) thermometer well, (e) electrolyte inlet, (f) gaseous product outlet, (g) reflux condenser, (h) cathode, (i) anode.

*Chapter 2*

# Fluorinating Agents

In order to introduce fluorine into organic compounds, the latter must be treated with hydrogen fluoride, fluorine, inorganic fluorides, or some organic fluorine derivatives capable of passing fluorine into them. Only a few fluorinating agents have to be prepared in the laboratory; the majority are now commercially available. The main American suppliers of the most important inorganic fluorinating agents are listed in Table 4, the prices in Table 5.

**Table 4. Codes, Names, and Addresses of the Main Suppliers of Inorganic Fluorides**[a]

| Code | Name of firm | Address |
|---|---|---|
| ACE | Aceto Chemical Company (Pfaltz & Bauer) | 126–04 Northern Blvd., Flushing, N.Y. 11368 |
| AIC | Allied Chemical Corporation (Baker & Adamson) | P.O. Box 80, Morristown, N.J. 07960 |
| ALF | Alfa Inorganics, Inc. | P.O. Box 159, Beverly, Mass. 01915 |
| APC | Air Product and Chemicals, Inc. | 733 West Broad St., Emmaus, Pa. 18049 |
| BKC | J. T. Baker Chemical Company | Phillipsburg, N.J. 08865 |
| CIC | City Chemical Corporation | 132 W. 22nd St., New York, N.Y. 10011 |
| CPL | Chemical Procurement Laboratories, Inc. | 18–17 130th St., College Point, N.Y. 11356 |
| DFG | D. F. Goldsmith Chemical and Metal Corp. | 909 Pitner Avenue, Evanston, Ill. 60602 |
| GSA | Gallard-Schlesinger Chemical Mfg. Corp. | 584 Mineola Ave., Carle Pl., Long Island, N.Y. 11514 |
| KNK | K&K Laboratories, Inc. | 121 Express St., Plainview, N.Y. 11803 |
| MAT | Matheson Gas Products | P.O. Box 85, East Rutherford, N.J. 07073 |
| MCB | Matheson, Coleman and Bell | 2909 Highland Ave., Norwood, Ohio 45212 |

Table 4 (Continued)

| Code | Name of firm | Address |
|---|---|---|
| OZM | Ozark-Mahoning Company | 1870 South Boulder, Tulsa, Okla. 74119 |
| PCR | Peninsular Chemresearch (Calgon Corporation) | P.O. Box 1466, Gainesville, Fla. 32601 |
| ROC | Research Organic/Inorganic Chemical Corp. | 11686 Sheldon St., Sun Valley, Calif. 91352 |
| VLO | Var-Lac-Oil Chemical Company | 666 South Front St., Elizabeth, N.J. 07202 |

[a] Chemical Sources, 1969 Edition, Directories Publishing Company, Flemington, N.J.

Table 5. Commercially Available Inorganic Fluorinating Agents (Prices in $)

| Compound | ACE | AIC (1969) | ALF (1969) | APC (1968) | BKC |
|---|---|---|---|---|---|
| HF | — | — | — | 2.50/lb, 10/0.5 lb[a] | 2.20/0.8 lb |
| NaF | 9/200 g | 2.81/lb | 37.20/100 | — | 3.23/lb |
| KF | — | 2.76/lb | — | — | — |
| $KSO_2F$ | 11/100 g | — | — | — | — |
| RbF | 5/10 g | — | 48/11 g | — | — |
| CsF | 10/10 g | — | 29/100 g | — | — |
| AgF | — | 55/lb | 52/100 g | — | — |
| $AgF_2$ | 15/50 g | 53/lb | 52/100 g | — | — |
| $Hg_2F_2$ | 7/10 g | — | — | — | — |
| $HgF_2$ | 9/20 g | 40/lb | 32/100 g | — | — |
| $BF_3$ | 27/150 g | 0.90/lb | — | 2.40/lb, 12/0.5 lb[a] | 3/0.45 lb |
| $BF_3 \cdot Et_2O$ | — | 1.34/lb | — | — | 3.90/kg |
| $NaBF_4$ | — | 1.25/lb | 7/lb | — | — |
| TlF | 12/50 g | — | 38/100 g | — | — |
| $Na_2SiF_6$ | 9/100 g | — | 4.20/lb | — | 8.40/lb |
| $AsF_3$ | 9/100 g | — | 8/100 g | — | — |
| $SbF_3$ | 7/100 g | 4.50/lb | 30/kg | — | — |
| $SbF_5$ | — | 25/lb | 150/2 lb | — | — |
| $SF_4$ | 135/lb | — | — | 75/lb[a] | 100/1.1 lb |
| $FSO_3H$ | 24/400 g | 8/0.7 kg | — | — | — |
| $F_2$ | — | 16/lb | — | 25/lb | 50/lb |
| $ClF_3$ | — | 4.50/lb | — | 10.80/lb | 32/1.6 lb |
| $FClO_3$ | — | — | — | — | — |
| $BrF_3$ | 48/450 | 3.60/lb | — | 12/lb, 25/lb[a] | 53.50/2.5 lb |
| $IF_5$ | 60/lb | 16/lb | — | 40/lb, 75/lb[a] | 30/lb |
| $MnF_3$ | 24/100 g | 30/lb | 24/100 g | — | — |
| $CoF_3$ | 9/200 g | 11.10/lb | 7.40/100 | — | — |

Table 5 (Continued)

| Compound | CIC (1964) | CPL (1969) | DFG | GSA (1969) | KNK |
|---|---|---|---|---|---|
| HF | — | — | — | — | — |
| NaF | 0.7–3.29/lb | — | — | — | 5.50/100 g |
| KF | 1.85/lb | — | — | — | — |
| $KSO_2F$ | 22/lb | 18/100 g | — | — | 14.50/100 g |
| RbF | 26/100 | 15/10 g | 45/100 g | — | 40/100 g |
| CsF | 17.50/100 g | 30/100 g | 50/100 g | — | 32.50/100 g |
| AgF | — | 100/100 g | 27.50/100 g | 125/lb | 34.50/100 g |
| $AgF_2$ | 66.50/lb | — | — | 125/lb | 34.50/100 g |
| $Hg_2F_2$ | 19.50/lb | 40/100 g | — | — | 9/10 g |
| $HgF_2$ | 89/lb | 33/100 g | — | 100/lb | 34.50/100 g |
| $BF_3$ | 11/2 lb | 85/2 lb | — | 100/lb | 32.50/150 g |
| $BF_3 \cdot Et_2O$ | 1.85/lb | — | — | — | — |
| $NaBF_4$ | 1.20/lb | — | — | — | — |
| TlF | — | 30/100 g | 17/100 g | — | 29/100 g |
| $Na_2SiF_6$ | 0.85–8.15/lb | — | — | — | 11/100 g |
| $AsF_3$ | 12/lb | 35/lb | 12/lb | 30/lb | 50/kg |
| $SbF_3$ | 5.50/lb | 10/100 g | 7/lb | — | 8.50/100 g |
| $SbF_5$ | 28/lb | 60/lb | 20/100 g | 100/lb | 67.50/500 g |
| $SF_4$ | 55.20/lb | 190/lb | — | — | 157.50/lb |
| $FSO_3H$ | 4.20/100 g | — | — | — | 32.50/800 g |
| $F_2$ | — | — | — | — | — |
| $ClF_3$ | 48/500 g | 50/lb | — | — | 45/500 g |
| $FClO_3$ | — | — | — | — | — |
| $BrF_3$ | — | — | — | — | 55/lb |
| $IF_5$ | — | 80/lb | — | 100/lb | 67.50/lb |
| $MnF_3$ | 68/lb | 50/100 g | — | — | 29/100 g |
| $CoF_3$ | 18.10/lb | 15/100 g | — | — | 14/100 g |

| Compound | MAT (1969) | MCB (1969) | OZM (1969) | PCR | ROC (1969) | VLO (1960) |
|---|---|---|---|---|---|---|
| HF | 2.50/0.75 lb | — | — | — | — | — |
| NaF | — | 3.33/lb | 6/2 lb | — | 3.50/lb | — |
| KF | — | 3/lb | 5/lb | — | 5/lb | — |
| $KSO_2F$ | — | — | — | 12/100 g, 25/500 g | 11/100 g, 40/lb | — |
| RbF | — | 19.80/25 g | — | — | 12/25 g, 45/100 g | 20/10 g |
| CsF | — | 13.10/25 g | 15/100 g | 15/100 g | 15/100 g, 55/lb | 5/10 g |
| AgF | — | 8/10 g | 20/100 g | — | 32/100 g, 95/lb | — |
| $AgF_2$ | — | — | — | — | 50/100 g | — |
| $Hg_2F_2$ | — | — | — | — | — | 18/10 g |
| $HgF_2$ | — | — | — | — | 30/100 g, 95/lb | 18/25 g |

Table 5 (Continued)

| Compound | MAT (1969) | MCB (1969) | OZM (1969) | PCR | ROC (1969) | VLO (1969) |
|---|---|---|---|---|---|---|
| $BF_3$ | 6.35/2 lb | — | — | — | 30/2 lb | 38.50/150 g |
| $BF_3 \cdot Et_2O$ | — | 4.10/kg | — | — | — | 12/1 kg |
| $NaBF_4$ | — | 2/lb | 5/lb | — | 6/lb | — |
| TlF | — | — | — | — | 35/100 g | 34/100 g |
| $Na_2SiF_6$ | — | 9/lb | 5/2 lb | — | 3.50/lb | — |
| $AsF_3$ | — | — | 5/lb | 8/100 g, 16/500 g | 30/lb | 20/lb |
| $SbF_3$ | — | 5/0.25 lb | 5/lb | 8/100 g, 15/500 g | 12/lb | 10/lb |
| $SbF_5$ | — | — | 25/200 g | 15/100 g | 50/lb | 45/200 g |
| $SF_4$ | 110/lb | — | — | 60/100 g, 180/500 g | 125/lb | — |
| $FSO_3H$ | — | — | — | — | 18/lb | — |
| $F_2$ | 44/0.5 lb | — | — | — | — | — |
| $ClF_3$ | 22/lb | — | 27/lb | — | — | — |
| $FClO_3$ | — | — | 50/0.5 lb | — | — | — |
| $BrF_3$ | 23/lb | — | 30/lb | — | 35/lb | — |
| $IF_5$ | 36.50/lb | — | — | 50/lb | 60/lb | — |
| $MnF_3$ | — | — | — | — | 24/100 g | 28/100 g |
| $CoF_3$ | — | — | — | 10/100 g, 20/500 g | 15/lb | 10/100 g |

[a] Includes price of lecture bottle.

## HYDROGEN FLUORIDE

Hydrogen fluoride is obtained by decomposing acid-grade (97%) fluorspar with concentrated sulfuric acid in mild-steel rotary kilns at temperatures up to 350°C. The gaseous hydrogen fluoride is either condensed and redistilled, or absorbed in azeotropic aqueous hydrofluoric acid until saturation in the cold (70–80%). Distillation of such a solution and redistillation of the first distillate gives anhydrous hydrogen fluoride of 99.9% purity, satisfactory for most purposes **[10]**.

It is delivered in steel cylinders from which it can be drawn as a liquid (by draining from the bottom of the cooled cylinder), or as a gas (from the top of the cylinder, heated above 20°C). The gaseous hydrogen fluoride may be led through a plastic or metal condenser attached to the cylinder and collected as a liquid.

For special purposes, e.g., hydrogenation, hydrogen fluoride must be purified in order to remove small amounts of sulfur dioxide present. Hydrogen fluoride (16.5 kg) containing 0.16% of sulfur dioxide is shaken with 900 g of manganese dioxide for 8 hr at 80°C and 6.5 atm in an autoclave. Distillation after discarding the first 200 g of the distillate gives 15.75 kg of pure hydrogen fluoride free of sulfur dioxide and suitable for catalytic hydrogenations **[11]**.

Aqueous solution of hydrogen fluoride—hydrofluoric acid—is available in concentrations of 40%, 48%, and 65–70%. An azeotropic mixture of hydrogen fluoride and water boils at 112.0°C at 750.2 atm and contains 38.26% of hydrogen fluoride. The vapor–liquid diagram of aqueous hydrofluoric acid shows how easily hydrogen fluoride can be obtained by distillation (Fig. 3) **[12]**.

## FLUORINE

Elemental fluorine is produced by electrolysis of a salt KF·2HF at 100°C using a potential drop of 12–18 V and a current density of 5–10 A $dm^{-1}$ in steel cells with nickel or carbon anodes, steel cathodes, and steel diaphragms. The cell itself may function as the cathode. Lithium fluoride is added as melting-point depressant **[13]**.

Elemental fluorine is delivered in 98% purity in steel or stainless steel tanks under the pressure of 21–28 atm.

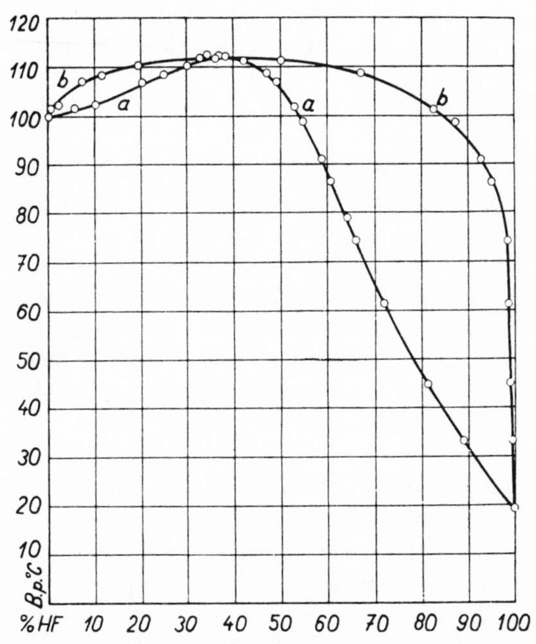


**Fig. 3.** Liquid–vapor diagram in water–hydrogen fluoride system: (a) liquid, (b) vapor (both in weight per cent).

**Table 6. Physical Properties of Fluorine and Hydrogen Fluoride**

| Property | Fluorine | Hydrogen fluoride | Vapor pressure of hydrogen fluoride | |
|---|---|---|---|---|
| | | | °C | mm Hg |
| M.P., °C | −218.0 | −83.36 | −40 | 53 |
| B.P., °C | −187.99 | 19.51 | −20 | 150 |
| Critical temperature, °C | −129.2 | 230.2 | 0 | 360 |
| Density at −60°C | — | 1.1660 | 20 | 780 |
| −30°C | — | 1.0735 | 40 | 1500 |
| 0°C | — | 1.0015 | 60 | 2600 |
| +30°C | — | 0.955 | 80 | 4400 |

Equation for the calculation of vapor pressure of hydrogen fluoride:

$$\log P_{\mathrm{mm}} = 8.38036 - \left\{\frac{1952.55}{[335.52 + t\ (^\circ\mathrm{C})]}\right\}$$

or

$$- 1.91173 - \left(\frac{918.24}{T}\right) + 3.21524 \log T$$

Physical constants of hydrogen fluoride and fluorine are listed in Table 6; the vapor pressure of hydrogen fluoride is shown in Fig. 4 (p. 17).

## INORGANIC FLUORIDES

Most of the fluorides used as fluorinating agents are now available commercially (Tables 4 and 5). Nevertheless, occasionally, need may arise for preparing some of them. Sometimes, it is more convenient to prepare an inorganic fluoride from available material than to await its delivery from a supplier. For this purpose, a guide to the preparation of the most useful fluorinating agents is given in Table 7.

A survey of physical constants and applications of individual inorganic fluorides in the preparation of organic fluorine compounds is shown in Table 8.

## ORGANIC FLUORIDES

The number of organic fluorinating agents is small but is gradually increasing. The use of aryl iodide difluorides **[32]** is now outdated, trifluoromethylhypofluorite is used only to a limited extent **[33]**, carbonyl chlorofluoride and carbonyl fluoride **[29,30]** are suitable for the preparation of fluoroformates, and phenylsulfur trifluoride **[28]** for the replacement of a carbonyl oxygen by fluorine. Diphenyltrifluorophosphorane, triphenyldi-

Table 7. Survey of the Preparation of Some Inorganic Fluorides (Starting Material in Parentheses)[a]

| Fluorinating agent | | | |
|---|---|---|---|
| Aqueous hydrofluoric acid (40–70%) | Anhydrous hydrogen fluoride | Elementary fluorine | Other reagents |
| $NaF$ ($NaOH$, $Na_2CO_3$) | $HgF_2$ ($HgO$, $HgCl_2$) | $AgF_2$ ($Ag$, $AgHal$) | $HgF_2$ ($Hg_2F_2+Cl_2$) |
| $KF$ ($KOH$, $K_2CO_3$) | $AsF_3$ ($AsCl_3$) **[19]** | $ClF_3$ ($Cl_2$) | $BF_3$ ($NaBF_4+H_2SO_4$) **[22]** |
| $RbF$ ($Rb_2CO_3$) | ($As_2O_3$) **[20]** | $BrF_3$ ($Br_2$) | ($BF_3 \cdot Et_2O \rightarrow BF_3 \cdot PhNH_2+H_2SO_4$) **[23]** |
| $CsF$ ($Cs_2CO_3$) | $SbF_5$ ($SbCl_5$) | $IF_5$ ($I_2$) | $SF_4$ ($SCl_2+NaF$) **[24,25]** |
| $AgF$ ($Ag_2O$) **[14]** | $FSO_3H$ ($ClSO_3H$) **[21]** | $MnF_3$ ($MnF_2$) | $KF \cdot SO_2$ ($KF+SO_2$) **[26]** |
| ($Ag_2CO_3$) **[15]** | | $CoF_3$ ($CoCl_2$, $Co_2O_3$) | $FClO_3$ ($KClO_4+FSO_3H$) **[27]** |
| $Hg_2F_2$ ($Hg_2CO_3$) **[16]** | | | $C_6H_5SF_3$ [$(C_6H_5S)_2+AgF_2$] **[28]** |
| $HBF_4$ ($H_3BO_3$) **[17]** | | | $COClF$ ($COCl_2+SbF_3$) **[29]** |
| $NaBF_4$ ($H_3BO_3$) | | | $COF_2$ ($COCl_2+SbF_3$) **[29]** |
| $TlF$ ($TlOCOH$) **[18]** | | | ($COCl_2+NaF$) **[30]** |
| $Na_2SiF_6$ ($SiO_2$) | | | $CHClFCF_2N(C_2H_5)_2$ **[31]** |
| $SbF_3$ ($Sb_2O_3$) | | | [$CClF{=}CF_2+NH(C_2H_5)_2$] |

[a] Since boron trifluoride etherate is common in the laboratory, a simple procedure of converting it to gaseous boron trifluoride may be useful **[23]**: Aniline (28 g, 0.3 mole) is added dropwise over 30 min to a stirred solution of 42.6 g (0.3 mole) of boron trifluoride etherate in 100 ml of absolute benzene cooled to 10°C. After stirring at room temperature for 30 more minutes, the precipitate is filtered with suction to give 47 g (97%) of a boron trifluoride–aniline complex. This, dried and heated with 120 ml of concentrated sulfuric acid with a free flame under a reflux condenser, gives 11–12 g (approximately 65%) of gaseous boron trifluoride.

**Table 8. Physical Properties of Inorganic Fluorides**

| Fluoride | Mol. wt. | M.P., °C | B.P., °C | $d/t$ (°C) | Solubility, g/100 g $H_2O/t$ | Applications in org. chem. | | | | |
|---|---|---|---|---|---|---|---|---|---|---|
| | | | | | | Addn. F | Repl. H | Repl. Hal | Repl. O | Repl. N |
| HF | 20.01 | −83.36 | 19.51 | 1.0015/10 | Unlimited | * | — | * | * | * |
| NaF | 41.99 | 992 | 1704 | 2.726 | 4/0, 4.22/18, 5/100 | — | — | * | — | — |
| KF | 58.10 | 858 | 1502 | 2.528 | 92.3/18 | * | — | * | * | — |
| RbF | 104.48 | 775 | 1408 | 3.202 | — | — | — | * | — | — |
| CsF | 151.91 | 703 | 1253 | 3.586 | 366/18 | — | — | * | — | — |
| AgF | 126.88 | 435 | ~1150 | 5.852/15.5 | 182/15.5, 205/108 | — | — | * | — | — |
| $AgF_2$ | 145.88 | ~690 | — | 4.57–4.78 | Decomp. | * | * | * | — | — |
| $Hg_2F_2$ | 439.22 | 570 | <650 decomp. | 8.73 | Decomp. | — | — | * | — | — |
| $HgF_2$ | 238.61 | 645 | Decomp. | 8.95 | Decomp. | * | — | * | — | — |
| $BF_3$ | 67.82 | −127.1 | −101.0 | 3.0 g/liter/20 | 106 ml | — | — | — | — | — |
| $BF_3{\cdot}(C_2H_5)_2O$ | 141.94 | −60.4 | 125.7 | 1.125 | — | — | — | — | * | — |
| $HBF_4$ | 87.83 | — | 130 decomp. | — | Unlimited | — | — | — | — | * |
| $NaBF_4$ | 109.81 | 384 | Decomp. | 2.47 | 108/26, 210/100 | — | — | — | — | * |
| TlF | 223.39 | 327 | 655 | — | 78.6/15 | — | — | * | — | — |
| $Na_2SiF_6$ | 188.07 | Decomp. | — | 2.679 | 0.652/17, 2.46/100 | — | — | * | — | * |
| $AsF_3$ | 131.91 | −5.97 | 57.13 | 2.666 | Decomp. | — | — | * | — | — |
| $SbF_3$ | 178.76 | 292 | 319 | 4.379 | 384.7/0, 563.6/30 | — | — | * | — | — |
| $SbF_5$ | 216.76 | 8.3 | 141 | 3.145/15.5, 2.993/22.7 | — | * | — | * | — | — |
| $SF_4$ | 108.01 | −121.0 | −40.4 | — | — | — | — | * | * | — |
| $FSO_3H$ | 100.02 | −87.3 | 165.5 | 1.743/15 | Decomp. | — | — | — | — | — |
| $ClF_3$ | 92.46 | −76.34 | 11.75 | 1.88 | — | * | * | * | — | — |
| $FClO_3$ | 102.46 | −147.74 | −46.67 | — | — | — | * | — | — | — |
| $BrF_3$ | 136.92 | 8.77 | 125.75 | 2.79/27 | — | * | * | * | — | — |
| $IF_5$ | 221.91 | 9.43 | 100.5 | 4.31/20 | — | * | — | * | — | — |
| $MnF_3$ | 111.94 | — | — | 3.54 | Decomp. | * | * | * | — | — |
| $CoF_3$ | 115.94 | 1200 | 1400 | 3.88 | Decomp. | * | * | * | — | — |

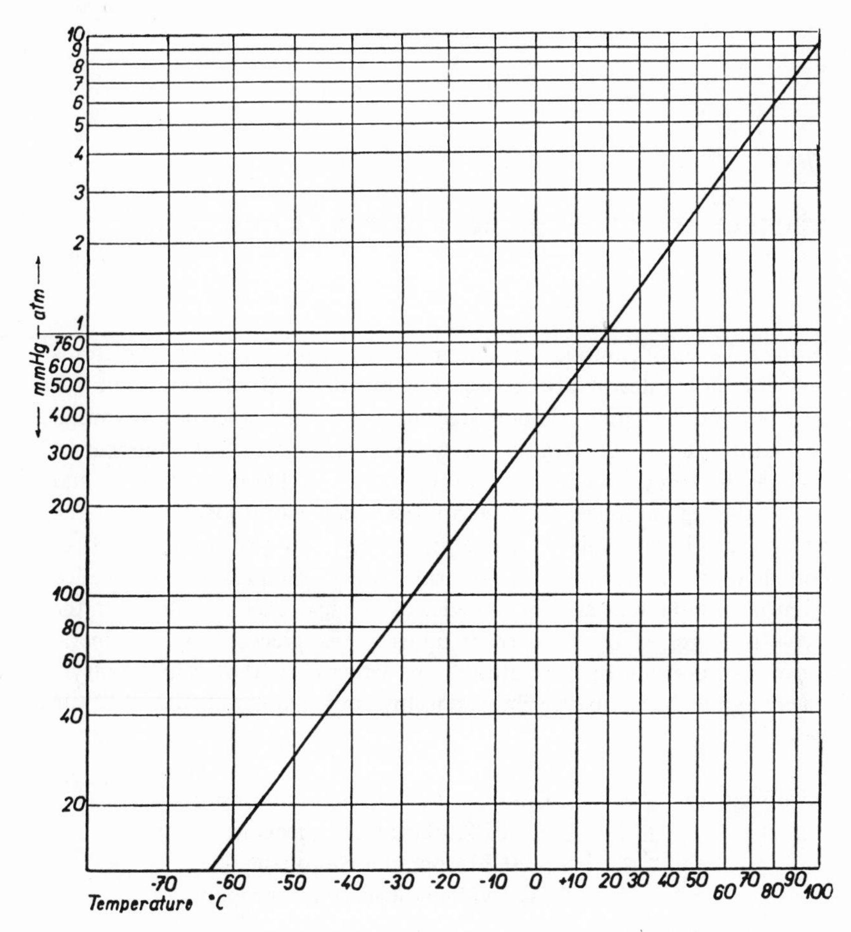


**Fig. 4.** Dependence of the vapor pressure of anhydrous hydrogen fluoride on temperature.

fluorophosphorane **[34]**, and especially chlorotrifluorotriethylamine **[31]** are favorite agents for the replacement of a hydroxyl group by fluorine.

The preparation of organic fluorinating agents is listed in Table 7; their applications will be given in the appropriate places later.

*Chapter 3*

# Nomenclature of Organic Fluorine Compounds

The nomenclature of compounds containing one or just a few fluorine atoms follows common rules involving use of numerals or Greek letters for the designation of the position of fluorine atoms in organic molecules. A special nomenclature had to be created for compounds in which the number of fluorine atoms prevails over that of hydrogen or halogen* atoms, and consequently, regular names would have been cumbersome **[35,36]**. The same difficulty arose in compounds where all of the hydrogen atoms bound to carbon had been replaced by fluorine. Such compounds are called "perfluoro" compounds, and the presence of any hydrogen or halogen in the molecule is expressed by the corresponding prefix preceded by a numeral indicating the position of these atoms. The letter H or the prefix "hydryl" (not hydro) stands for hydrogen. The following examples should illustrate the practice:

$CHF_2CClFCF_3$ — 2-Chloro-1,1,2,3,3,3,-hexafluoropropane,
1H,2-Chlorohexafluoropropane,
1H,2-Chloroperfluoropropane,
1-Hydryl-2-chloroperfluoropropane

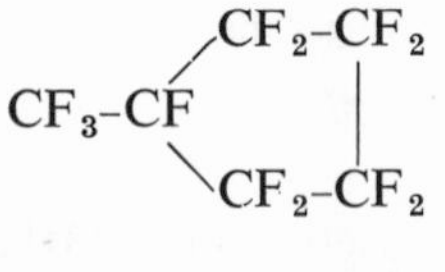

$$\begin{array}{l} \quad\quad\quad CF_2\text{-}CF_2 \\ CF_3\text{-}CF \quad\quad\;\; | \\ \quad\quad\quad CF_2\text{-}CF_2 \end{array}$$ — Perfluoro(methylcyclopentane)

The prefix "perfluoro" means the replacement of all hydrogen atoms bound to carbon atoms of the carbon chain. If hydrogen atoms in functional groups are also replaced by fluorine, the name must be modified accordingly, as in the following examples:

---

*Throughout the book, the term "halogen" will be applied to chlorine, bromine, and iodine, but in general not to fluorine, for practical reasons.

| | |
|---|---|
| $C_3F_7CHO$ | Perfluorobutyraldehyde |
| $C_2F_5COOH$ | Perfluoropropionic acid |
| $C_2F_5COOF$ | Perfluoropropionylhypofluorite |
| $C_6F_5NH_2$ | Perfluoroaniline |
| $C_4F_9NF_2$ | 1-Difluoroaminoperfluorobutane |
| $CF_3SH$ | Perfluoromethylmercaptan |
| $CF_3SF$ | Perfluoromethanesulfenyl fluoride |

Fluorinated derivatives of tetracovalent and hexacovalent sulfur represent an especially difficult nomenclatural problem:

$C_6H_5SF_5$ Phenylsulfurpentafluoride, pentafluorosulfanylbenzene

In the Fluorine Chemistry series edited by J. H. Simons, a special nomenclature for "perfluoro" compounds was used. The prefix "perfluoro" is abbreviated to "for," which is inserted into the chemical name before the ending expressing the type of compound. Fortunately, this chemical dialect did not survive.

| | |
|---|---|
| $CF_3CF{=}CF_2$ | Propforene |
| $CF_3C_6F_4CF_3$ | Dimethforylbenzforene |

The nomenclature of phosphorus–fluorine compounds can be exemplified as follows (both old and new nomenclature are used for derivatives of phosphorus acids):

| | |
|---|---|
| $CF_3PF_2$ (derivative of $PH_3$) | Trifluoromethyldifluorophosphine |
| $(CF_3)_2PF_3$ (derivative of $PH_5$) | Bis(trifluoromethyl)trifluorophosphorane |
| $(HO)_2PF$ (derivative of $H_3PO_3$) | Fluorophosphinic acid, phosphorofluoridous acid |
| $HOP(O)F_2$ (derivative of $H_3PO_4$) | Difluorophosphonic acid, phosphorodifluoridic acid |

Part of the nomenclature of fluorine compounds also involves the commercial designation of refrigerants and propellants which became very common especially in technical literature. This consists of a trade name of the product, such as Freon, Frigen, Isceon, etc., or simply *F* (for fluorocarbon) and of a numerical symbol expressing the chemical composition of the compound according to the following rules (Table 9):

1. First number (omitted when zero): number of carbon atoms −1.

2. Second number: number of hydrogen atoms +1.
3. Third number: number of fluorine atoms.
4. Number of chlorine atoms is not given.
5. Number of bromine atoms is preceded by B.
6. Numerical symbol of cyclic compounds is preceded by C.
7. Less common isomer is designated by *a* following the numerical symbol.

**Table 9. Nomenclature of Refrigerants and Propellants**

| | |
|---|---|
| *F* 11 | $CCl_3F$ |
| *F* 12 | $CCl_2F_2$ |
| *F* 12B2 | $CBr_2F_2$ |
| *F* 13B1 | $CBrF_3$ |
| *F* 22 | $CHClF_2$ |
| *F* 113 | $CCl_2FCClF_2$ |
| *F* 113a | $CCl_3CF_3$ |
| *F* 114 | $CClF_2CClF_2$ |
| *F* 114B2 | $CBrF_2CBrF_2$ |
| *F* C318 | $CF_2{-}CF_2$ (ring) / $CF_2{-}CF_2$ |

*Chapter 4*

# Introduction of Fluorine into Organic Compounds

The large number of methods for introducing fluorine into organic compounds may be subdivided into addition of hydrogen fluoride, addition of fluorine, addition of halogen fluorides across double and triple bonds, and substitution of fluorine for hydrogen, halogen, oxygen, and nitrogen.

## ADDITION OF HYDROGEN FLUORIDE TO OLEFINS

Olefinic hydrocarbons tend to polymerize in contact with anhydrous hydrogen fluoride. This circumstance decreases the general applicability and reliability of this reaction. Nevertheless, some aliphatic and alicyclic monofluorides were successfully prepared under special conditions such as low or appropriate temperature, excess of hydrogen fluoride, proper contact time, etc. **[37]**.

$$(1) \qquad CH_2{=}CH{-}CH_3 \xrightarrow[0°C,\ 3\ atm]{HF} \underset{60\%}{CH_3{-}CHF{-}CH_3} \qquad [37]$$

Olefins with halogen atoms linked to double-bond carbons show much less tendency to polymerize. Here, another side reaction decreases the yields of monofluorides: concomitant replacement of halogen atoms by hydrogen. This reaction can be cut down if not too high a temperature is used **[38]**, or a diluent is applied **[39]**. On the other hand, energetic conditions in such reactions represent a good way to obtain geminal polyfluorides.

$$(2) \qquad CCl_2{=}CHCH_2CH_3 \xrightarrow{HF} CCl_2FCH_2CH_2CH_3 + CClF_2CH_2CH_2CH_3 \qquad [38]$$

| Conditions | $CCl_2FCH_2CH_2CH_3$ | $CClF_2CH_2CH_2CH_3$ |
|---|---|---|
| 65°C, 18 hr | 33% | 5.3% |
| 100°C, 6 hr | 10.3% | 49.6% |

(3) [39]

$$\text{1-chlorocyclohexene} \xrightarrow[1.5\ hr]{HF,\ Et_2O} \text{1-chloro-1-fluorocyclohexane (96\%)}$$

$$\text{1-chlorocyclohexene} \xrightarrow[1\ hr]{HF(g)} \text{1-chloro-1-fluorocyclohexane (27.6\%)} + \text{1,1-difluorocyclohexane (43\%)}$$

Olefins having double bonds flanked by halogens and/or fluorine atoms require energetic conditions for the reaction to occur. Even so, addition of hydrogen fluoride does not take place in fluorohalo-olefins in which fluoride is attached to the internal carbon of a double bond **[40]**:

(4) [40]

HF added

$CF_2{=}CH{-}CH_3$, $CF_2{=}CH{-}CF_3$

HF not added

$CHCl{=}CF{-}CCl_3$, $CH_2{=}CF{-}CCl_2F$, $CH_2{=}CF{-}CClF_2$, $CH_2{=}CF{-}CF_3$

With perfluoro-olefins, electrophilic addition of hydrogen fluoride fails because of the drainage of $\pi$-electrons by fluorine atoms. By the same token, nucleophilic addition of fluoride ions across such a bond is much more favored and leads ultimately to the same products. This is carried out using ammonium or potassium fluoride in protic solvents **[41]**, silver fluoride in hydrogen fluoride **[42]**, or silver fluoride in acetonitrile followed by hydrolysis **[43]**.

(5)

$$CF_2{=}CF{-}CF_3 \longleftrightarrow \overset{\oplus}{C}F_2{-}\overset{\ominus}{\underset{-}{C}}F{-}CF_3 \longrightarrow CF_3{-}CHF{-}CF_3$$

[41] $Et_4NF$, $CHCl_3$, 20°C: 84%

[41] KF, $HCONH_2$, 25°C: 59%

[42] 10% AgF in HF, 125°C: 47%

[43] 1: AgF, MeCN, 25°C; 2: $H_2O$: 95%

## ADDITION OF HYDROGEN FLUORIDE TO ACETYLENES AND OTHER UNSATURATED SYSTEMS

One or two moles of hydrogen fluoride may be added to acetylenes. Special precautions must be taken for stopping the reaction at the stage of monofluoro-olefins, as in the preparation of vinyl fluoride **[44]**:

(6) [44]

$$CH{\equiv}CH \xrightarrow[97\text{-}104°C]{HF,\ HgCl_2,\ BaCl_2/C} \underset{82\%}{CH_2{=}CHF} + \underset{4\%}{CH_3{-}CHF_2}$$

Addition of two moles of hydrogen fluoride across a triple bond is easy **[45]**, and is advantageously carried out in solvents forming oxonium salts with hydrogen fluoride such as ethers and acetone **[46]**.

(7) $CH{\equiv}C{-}R \xrightarrow[-70^\circ \to 0^\circ \to 15^\circ C]{2HF} CH_3{-}CF_2{-}R$ 46–76% [45]

$CH{\equiv}C{-}R \xrightarrow[CH_3COCH_3]{5HF} CH_3{-}CF_2{-}R$ 85–90% [46]

In reactions of hydrogen fluoride with compounds having carbon–nitrogen bonds, proton joins the nitrogen atom **[47,48]**.

(8) $CF_2{=}N{-}CF_3 \xrightarrow[150^\circ C]{HF} CF_3{-}NH{-}CF_3$ 89% [47]

(9) $C_6H_5{-}N{=}C{=}O \xrightarrow[-80^\circ \to 20^\circ C]{HF,\ Et_2O} C_6H_5{-}NHCOF$ quant. [48]

Nucleophilic addition of a fluoride ion across a carbonyl bond occurs in the reaction of potassium or cesium fluoride with polychlorofluoro- or perfluoroketones. It leads to the salts of polychlorofluoro- or perfluoro-alcohols, very versatile and useful intermediates **[49,50]**.

(10) $CF_3{-}CO{-}CF_3 \xrightarrow[20^\circ C]{KF,\ diglyme} CF_3{-}CF(OK){-}CF_3$ >90% [49]

## ADDITION OF FLUORINE TO OLEFINS

The addition of fluorine across double bonds can be accomplished by various reagents. *Elemental fluorine* can be used for this purpose, but only with highly fluorinated substrates which are resistant to side reactions such as replacement of other atoms by fluorine. Even here, free-radical dimerization decreases the yields of the pure addition product **[51]**.

(11) $CF_3{-}CF{=}CF{-}CF_3 \xrightarrow[-77^\circ C,\ 7.5\,hr]{F_2,\ N_2(1:11)} CF_3{-}CF_2{-}\dot{C}F{-}CF_3$

$\xrightarrow{F_2} CF_3{-}CF_2{-}CF_2{-}CF_3$ 55%

$\rightarrow CF_3{-}CF_2{-}CF(CF_3){-}CF(CF_3){-}CF_2{-}CF_3$ 20% [51]

Another reagent suitable for adding fluorine across double bonds is *lead dioxide with hydrogen fluoride* **[52]**, or *lead dioxide with sulfur tetrafluoride* **[53]**.

(12) $$CCl_2{=}CCl_2 \xrightarrow[\text{Spont.}]{\text{HF, PbO}_2} CCl_2F{-}CCl_2F \quad 57\%$$ [52]

(13) [53]

$$\text{(cis) } ClHC{=}CHCl \xrightarrow[100°C,\ 2hr]{PbO_2,\ SF_4} \text{meso } CHClF{-}CHClF\ (21.5\%) + \text{DL } CHClF{-}CHClF\ (68.5\%)$$

$$\text{(trans) } ClHC{=}CHCl \xrightarrow[100°C,\ 2hr]{PbO_2,\ SF_4} \text{meso}\ (26.1\%) + \text{DL}\ (54.3\%)$$

Instead of lead dioxide, lead tetraacetate can be used in combination with hydrogen fluoride to achieve the addition of two fluorine atoms to olefinic compounds. The active reagent is *lead difluoride diacetate* **[54]**. Peculiar rearrangements occur during this reaction **[55,56]**. Similar rearrangements were also observed in the reaction of olefins with aryl iodide difluorides **[57]**.

(14) $$(C_6H_5)_2C{=}CH_2 \xrightarrow[0\text{-}20°C]{HF,\ Pb(OAc)_4} C_6H_5{-}CF_2{-}CH_2{-}C_6H_5 \quad 27\%$$ [55]

(15) [56]

Tri-O-acetyl glycal (CH₂OAc, OAc, AcO) $\xrightarrow[2.\ CH_3ONa]{1.\ HF,\ Pb(OAc)_4}$ 21.4%; ethyl glycoside (CH₂OAc, AcO, OEt) $\xrightarrow[2.\ CH_3ONa]{1.\ HF,\ Pb(OAc)_4}$ 67% → furanose product with $CH_2OH$, HO, OH, $CHF_2$

(16) $$C_6H_5{-}CH{=}CH{-}C_6H_5 \xrightarrow[0°C,\ 20\ min]{p\text{-}ClC_6H_4IF_2} (C_6H_5)_2CH{-}CHF_2 \quad 47\%$$ [57]

## ADDITION OF FLUORINE TO OTHER UNSATURATED SYSTEMS

Examples of the addition of fluorine across carbon–carbon triple bonds are very scarce. The addition of fluorine to aromatic systems is usually accompanied by concomitant replacement of hydrogen by fluorine regardless of whether elemental fluorine, high-valency fluorides, or electrochemical fluorination are used (pp. 29, 30).

Addition of fluorine across carbon–nitrogen bonds in nitriles, thiocyanates, isocyanates, and isothiocyanates can be carried out using numerous reagents **[58–69]**. The results are shown in the following schemes and in Table 10:

(17)

$$\text{[58]}\ CF_3NF_2 + (CF_3)_2NF \xleftarrow{F_2} HC{\equiv}N \xrightarrow{F_2} \text{[58]}$$

$$\text{[59]}\ (CF_3)_2NH \xleftarrow{IF_5} IC{\equiv}N \xrightarrow{IF_5} CF_3{-}N{=}N{-}CF_3\ \text{[59]}$$

$$\text{[62]}\ CF_3{-}N{=}CF_2 \xleftarrow[CuF_2]{HgF_2} ClC{\equiv}N \xrightarrow[AgF_2\ [61],\ MnF_3\ [62]]{NaF+Cl_2\ [60],\ AgF+Cl_2\ [60]}$$

(18) **[63]**

$$NaC{\equiv}N \xrightarrow{SF_4} (29\%)\ CF_3{-}N{=}SF_2\ (37\%) \xleftarrow{SF_4} BrC{\equiv}N$$

$$NaSC{\equiv}N \xrightarrow{SF_4} (69\%)\ CF_3{-}N{=}SF_2\ (25\text{–}30\%) \xleftarrow{SF_4} (HN{=}C{=}O)_3$$

Addition of fluorine across carbon–oxygen double bond in carbon monoxide or carbonyl fluoride leads to trifluoromethylhypofluoride **[70,71]**, addition across carbon–sulfur bonds leads to compounds containing tetra- and hexacovalent sulfur **[72]**, and addition to trivalent phosphorus compounds gives fluorinated phosphoranes **[73]**.

(19)

$$CO \xrightarrow[170°C]{F_2,\ AgF_2} CF_3OF\ (70\%) \xleftarrow[170°C]{F_2,\ AgF_2} FCOF \qquad [70,\ 71]$$

(20)

$$CS_2 \xrightarrow[electro]{HF,\ NaF} F_3S{-}CF_2{-}SF_3 + F_5S{-}CF_2{-}SF_5 + \underset{>90\%}{CF_3{-}SF_5} \qquad [72]$$

(21)

$$(C_6H_5)_3P \xrightarrow[\substack{50°C\ 100°C\ 150°C \\ 4hr\ \ 4hr\ \ 6hr}]{SF_4} \underset{(C_6H_5)_3PF_2}{69\%\ \ 67\%} \xleftarrow[\substack{50°C\ 100°C\ 150°C \\ 4hr\ \ 4hr\ \ 6hr}]{SF_4} (C_6H_5)_3PO \qquad [73]$$

## ADDITION OF HALOGEN FLUORIDES AND ORGANIC FLUORIDES TO OLEFINS

Halogen monofluorides are rather unstable, and hardly any case of addition across a double bond of ready-made halogen fluorides has been reported. There are, however, quite a few methods for adding fluorine and

**Table 10. Reactions of Carbon–Nitrogen Multiple Bonds with Fluorinating Agents**

| Compound | Fluorinating agent | | | | |
|---|---|---|---|---|---|
| | $F_2$ | $HgF_2$ | $AgF_2$ | $CoF_3$ | $SF_4$ |
| $CH_3—C{\equiv}N$ | $CF_3—CF_2—NF_2$<br>$CF_2{=}NF$<br>[64] | $CH_3—CF_2—NF_2$<br>$CH_3—CF{=}NF$<br>$CH_2{=}CF—NH_2$<br>$CH_2{=}C{=}NF$<br>[65,66] | | | $CH_2F—CF_2—N{=}SF_2$<br>[63] |
| $CF_3—C{\equiv}N$ | $CF_3—CF_2—NF_2$<br>$CF_3—CF{=}NF$<br>$C_2F_5N{=}NCF_3$<br>$C_2F_5N{=}NC_2F_5$<br>[67] | | $C_2F_5N{=}NC_2F_5$<br>[60] | | $CF_3—CF_2—N{=}SF_2$<br>[63] |
| $N{\equiv}C—C{\equiv}N$ | $F_2NCF_2CF_2NF_2$<br>$C_2F_5N{=}NCF_3$<br>$C_2F_5NFCF_3$<br>$CF_3—CF_2—NF_2$<br>[58] | $CF_3—N{=}CF_2$ [62] | $CF_2—CF_2$ (ring, $N{=}N$) [69] | $F_2NCF_2CF_2NF_2$<br>$CF_3—CF_2—NF_2$<br>$CF_3—NF—CF_3$<br>[62] | |
| $C_6H_5—N{=}C{=}O$ | | | | | $C_6H_5—N{=}SF_2$ [63] |
| $C_2H_5—N{=}C{=}S$ | | $C_2H_5—N{=}CF_2$ [68] | | | |

other halogens to olefins and other unsaturated compounds using various reagents.

A mixture of *anhydrous hydrogen fluoride and an "active-halogen" compound* such as organic hypochlorite or N-halogenoamide effects trans-addition subjected to polar influences **[74–76]**: fluorine joins the more positive, the other halogen the more negative end of the double bond (one exception is in steroids).

(22) $(CH_3)_2C{=}C(CH_3)_2$

→ HF, $(CH_3)_3C{-}OCl$; $(C_2H_5)_2O$, −78°C, 1hr; 0°C, 1hr → $(CH_3)_2CF{-}CCl(CH_3)_2$ 46%

→ HF, $\begin{matrix}CH_2{-}CO\\ CH_2{-}CO\end{matrix}{>}NBr$ → $(CH_3)_2CF{-}CBr(CH_3)_2$ 60%

→ HF, $\begin{matrix}CH_2{-}CO\\ CH_2{-}CO\end{matrix}{>}NI$; $(C_2H_5)_2O$ → $(CH_3)_2CF{-}CI(CH_3)_2$ 55%

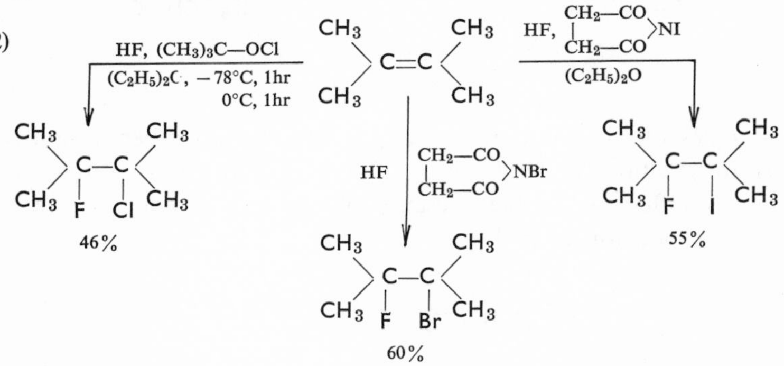

**[74]**

(23) Cyclohexene

→ HF, $CH_3CONHBr$; $(C_2H_5)_2O$, −80°C, 2hr; 0°C, 2hr → trans-1-bromo-2-fluorocyclohexane 42.5%

→ HF, $\begin{matrix}CH_2{-}CO\\ CH_2{-}CO\end{matrix}{>}NI$; $(C_2H_5)O_2$, −80°C, 2hr; 0°C, 1hr → trans-1-iodo-2-fluorocyclohexane 72.3%

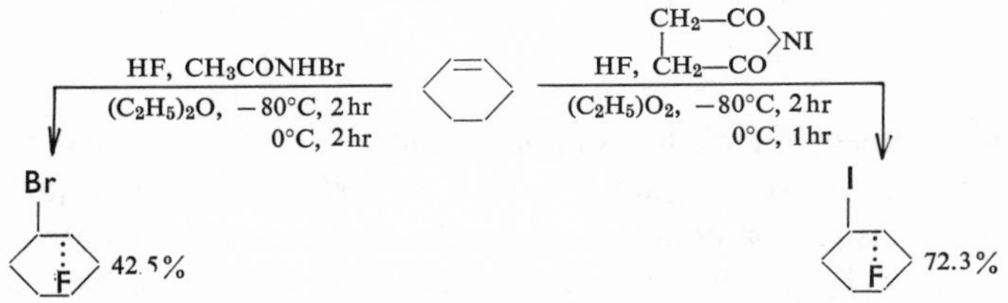

**[75]**

(24)
$$CH_2{=}CH{-}CO_2CH_3 \xrightarrow[CHCl_3,\ -30°C,\ 4hr]{HF,\ (CH_3)_2C{-}CO{-}NBr{-}CO{-}NBr} FCH_2{-}CHBr{-}CO_2CH_3\ 45\% + BrCH_2{-}CHF{-}CO_2CH_3\ \text{trace}$$

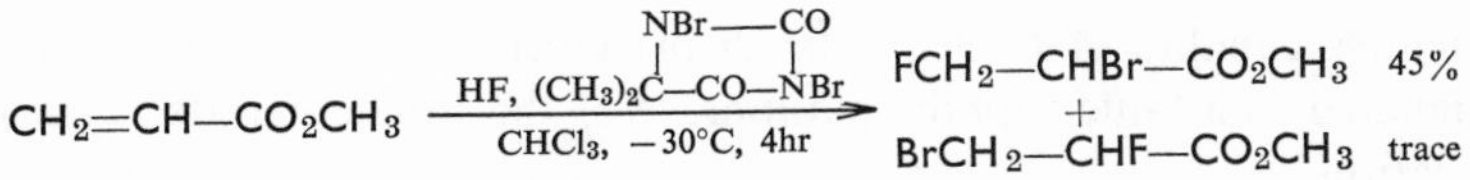

**[76]**

*Bromine trifluoride and bromine*, and *iodine pentafluoride and iodine*, also add fluorine and the other halogen across double bonds. The mechanism of the reaction is difficult to rationalize since the results are subject to reaction conditions and are partly contradictory **[77,78]**.

(25) $CF_2{=}CFCl$

→ $BrF_3$, $Br_2$; 20°C, 2 hr → $CF_3{-}CFClBr$ (13%) + $CF_2Br{-}CF_2Cl$ (73%)

→ $IF_5$, $2I_2$; 20°C, 2 hr → $CF_3{-}CFClI$ (37%) + $CF_2Cl{-}CF_2I$ (45%)

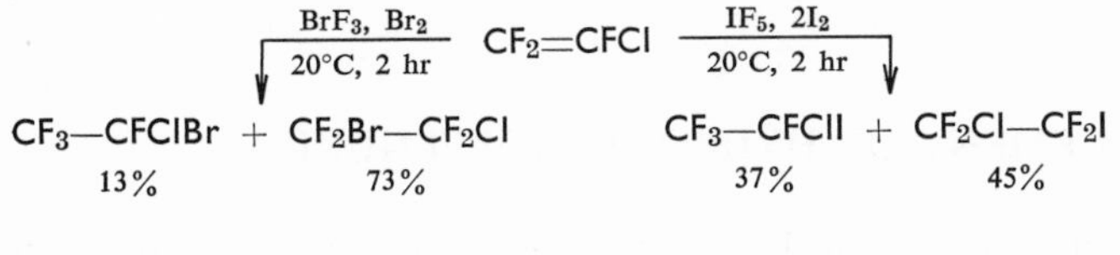

**[77]**

Another way of adding fluorine and iodine to olefins is the treatment of an olefin with a mixture of *potassium fluoride and iodine* in polar solvents. This reaction is limited to perfluoro-olefins which favor nucleophilic attack by fluorine. Consequently, the reaction is unidirectional **[79]**.

(26)

$$CF_3{-}CF{=}CF_2 \xrightarrow[150°C,\ 24\ hr]{IF,\ 2I_2\ [77]} CF_3{-}CFI{-}CF_3\ \ 99\%$$

$$CF_3{-}CF{=}CF_2 \xrightarrow[CH_3CN,\ 100°C,\ 10\ hr]{KF,\ I_2\ [79]} CF_3{-}CFI{-}CF_3\ \ 61\%$$

Examples of the addition of halogen fluorides to other unsaturated systems are reactions of perfluoroacyl fluorides or perfluoro ketones with chlorine monofluoride in the presence of cesium fluoride to give perfluoroalkyl hypochlorites **[80]**.

(27) $$CF_3{-}COF + ClF \xrightarrow[-20°C,\ 2\text{-}3\ hr]{CsF} CF_3{-}CF_2{-}OCl \quad \text{quant.}$$ **[80]**

(28) $$CF_3{-}CO{-}CF_3 + ClF \xrightarrow[-20°C,\ 2\text{-}3\ hr]{CsF} CF_3{-}CF(OCl){-}CF_3 \quad \text{quant.}$$ **[80]**

An extremely interesting example of addition is the reaction of *perfluoro-olefins with perfluoroacyl fluorides* **[81]** *or perfluoroaromatics* **[82]** in the presence of cesium or potassium fluoride, respectively. Evidently, fluorine ion converts the perfluoro-olefin to a perfluoroalkyl carbanion which displaces fluoride in acyl fluorides and strongly deactivated perfluoroaromatics capable of nucleophilic displacements. The reaction is suitably designated as a "nucleophilic paraphrase of the Friedel-Crafts acylation or alkylation."

(29)

$$CF_3{-}CF{=}CF_3 \xrightarrow[CH_3CN,\ 100°C]{CF_3COF,\ CsF\ [81]} CF_3{-}CF(CO{-}CF_3){-}CF_3\ \ 75\%$$

$$CF_3{-}CF{=}CF_3 \xrightarrow[(CH_2)_4SO_2,\ 130°C,\ 12\ hr]{KF,\ C_5F_5N\ [82]} CF_3{-}CF(C_5F_4N){-}CF_3\ \ 94\%$$

(the product of [82] is the 4-substituted tetrafluoropyridine: C attached to CF, FC=C, FC–CF, ring N)

## REPLACEMENT OF HYDROGEN BY FLUORINE

Substitution of fluorine for hydrogen is a nonselective reaction. Rarely can individual atoms of hydrogen be replaced; usually, polyfluoro and perfluoro compounds result from the action of elemental fluorine, high-valency metal fluorides, or an electrochemical fluorination process. One of the reasons is the high reaction heat of fluorination reactions, especially in the case of elemental fluorine. When high-valency metal fluorides are used, the reaction heat is reduced to about half, but even so, undesirable by-products due to carbon–carbon cleavage are always formed.

$$C{-}H + F_2 \longrightarrow C{-}F + HF \quad \Delta H_{298}{-}104 \text{ kcal/mole} \tag{30}$$
$$2CoF_2 + F_2 \longrightarrow 2CoF_3 \quad \Delta H_{473}{-}52 \text{ kcal/mole}$$

Except for the reactions of some organic compounds with perchloryl fluoride (see p. 30), instances of replacement of individual hydrogen atoms by fluorine are rare and lack in practical importance. On the other hand, reactions of organic compounds with *elemental fluorine* and high-valency metal fluorides and the electrochemical fluorination process are well suited for the preparation of polyfluoro and especially perfluoro compounds. Non-catalytic fluorination gives inferior yields to the catalytic process carried out over copper coated with silver or gold. These metals transiently form fluorides which pass fluorine onto the organic material **[83–85]**.

$$C_6H_6 \xrightarrow[265^\circ C]{[83]\; F_2,\, N_2;\, Ag{-}Cu} 58\%\; C_6F_{12} \tag{31}$$
$$C_6H_6 \xrightarrow[250\text{-}280^\circ C]{[84]\; F_2,\, N_2;\, Au{-}Cu} 51\%\; C_6F_{12}$$

Alicyclic hydrocarbons give generally better yields of perfluoro derivatives than the corresponding aromatic hydrocarbons. Comparison of the results with fluorination with fluorine over a gold catalyst is shown in Table 11 **[85]**.

**Table 11. Comparison of the Results of Fluorination of Alicyclic and Aromatic Hydrocarbons with Fluorine over a Gold Catalyst at 250–280°C [85]**

| Starting compound | Yield, % | Product | Yield, % | Starting compound |
|---|---|---|---|---|
| Ethylbenzene | 9.6 | Perfluoromethyl-cyclohexane | 21 | Ethylcyclohexane |
| *m*-Xylene | 4.5 | Perfluoro-1,3-dimethylcyclohexane | 15.4 | 1.3-Dimethylcyclo-hexane |
| Mesitylene | 5.9 | Perfluoro-1,3,5-trimethylcyclohexane | 11.2 | 1,3,5-Trimethyl-cyclohexane |
| Tetralin | 11.4 | Perfluorodecalin | 18.8 | Decalin |

Fluorination with elemental fluorine gave way to reactions of organic compounds with *high-valency metal fluorides*, which generally give higher yields. Cobalt trifluoride is now used almost exclusively, having supplanted the more expensive silver difluoride, also capable of the same reactions **[86,87]**.

(32) $$C_7H_{16} \xrightarrow[150\text{-}165^\circ C,\ 275\text{-}300^\circ C]{CoF_3} C_7F_{16}\ \ 91\%$$ **[86]**

(33) $$H_3C\text{—}C_6H_4\text{—}CH(CH_3)_2 \xrightarrow[260\text{-}280^\circ C]{CoF_3} F_3C\text{—}(C_6F_{10})\text{—}CF(CF_3)_2\ \ 50.5\%$$ **[87]**

The *electrochemical fluorination process* is especially suitable for converting compounds which are soluble in anhydrous hydrogen fluoride to perfluoro derivatives. The most useful application so far is the preparation of perfluoroacyl fluorides, which yield perfluorocarboxylic acids or their derivatives. Acyl fluorides give higher yields than acyl chlorides or anhydrides **[88,89]**.

(34)
[88] $(CH_3CO)_2O$
[89] $CH_3COF$
$$\xrightarrow{HF,\ NaF,\ 5.4\ V,\ 0.02\ A\ dm^{-2}} CF_3\text{—}COF \xrightarrow{H_2O} CF_3CO_2H\ \ 71\%$$

Exceptionally, individual hydrogen atoms can also be replaced by fluorine. For this purpose, *perchloryl fluoride* is the agent of choice. It reacts with nitro compounds and $\beta$-dicarbonyl compounds having "acidic hydrogens" and with enol ethers and enamines, and substitutes fluorine for $\alpha$-hydrogen atoms. The work with perchloryl fluoride requires certain precautions. Exit gases of reactions with perchloryl fluoride should never be cooled below 40°C since liquefied perchloryl fluoride tends to explode in contact with organic compounds. The usefulness of this reagent is documented in the following examples **[90–94]**:

(35) $$C_6H_5CH(NO_2)CH_3 \xrightarrow[2.\ FClO_3]{1.\ NaH,\ THF} C_6H_5CF(NO_2)CH_3\ \ 82\%$$ **[90]**

(36) $$CH_3COCH_2COCH_3 \xrightarrow[2.\ FClO_3,\ -10\text{–}0^\circ C]{1.\ MeONa,\ EtOH} CH_3COCF_2COCH_3\ \ 77\%$$ **[91]**

(37)

$$CH_2(CO_2C_2H_5)_2 \xrightarrow[2.\ FClO_3]{1.\ Na,\ C_7H_8} CHF(CO_2C_2H_5)_2 \xrightarrow[2.\ FClO_3]{1.\ Na,\ C_7H_8} CF_2(CO_2C_2H_5)_2$$ **[92]**

(38) [93]

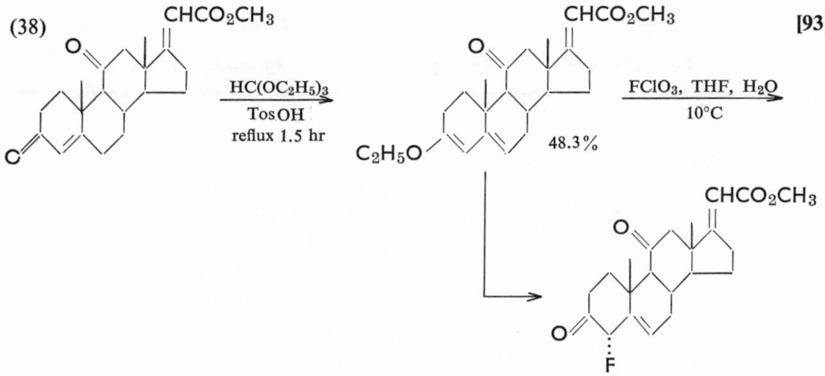

$CHCO_2CH_3$ → ($HC(OC_2H_5)_3$, TosOH, reflux 1.5 hr) → $C_2H_5O$ ... $CHCO_2CH_3$, 48.3% → ($FClO_3$, THF, $H_2O$, 10°C) → $CHCO_2CH_3$ ... F

(39) [94]

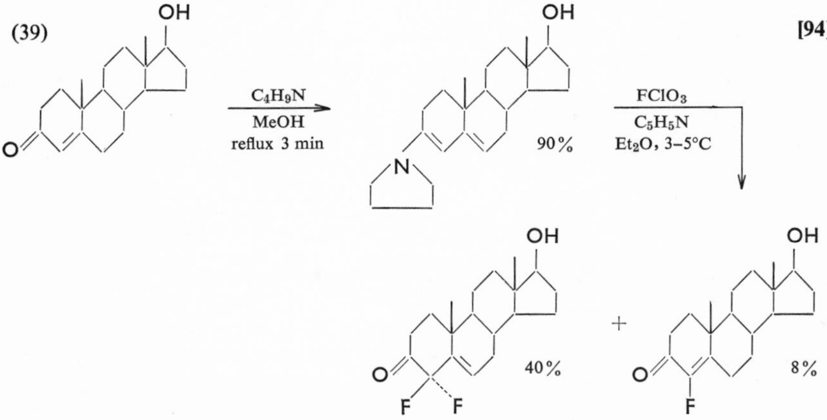

OH → ($C_4H_9N$, MeOH, reflux 3 min) → OH, N, 90% → ($FClO_3$, $C_5H_5N$, $Et_2O$, 3–5°C) → OH, F F, 40% + OH, F, 8%

## REPLACEMENT OF HALOGENS BY FLUORINE

Substitution of fluorine for other halogen atoms* is the oldest method of introducing fluorine into organic molecules, and is still of great importance. Originally, antimony trifluoride was the most useful reagent for this purpose. In time, it was replaced by the cheaper and more universal potassium fluoride. This compound is now also gradually supplanting other metal fluorides that were used for more selective purposes, such as silver fluoride and mercury fluorides. In industry, hydrogen fluoride in a non-catalytic, or better still, liquid- or vapor-phase catalytic fluorination process, is generally employed. A survey of the applicability of the main fluorides for the replacement of halogen atoms by fluorine in organic molecules is given in Table 12.

*As noted earlier, it is of advantage in this book and in fluorine chemistry to use the term halogen to mean chlorine, bromine, and iodine, but not fluorine.

**Table 12. Conversion of Organic Halides to Organic Fluorides[a]**

| Type of halogen bond | Reagent | | | | |
|---|---|---|---|---|---|
| | HF | $SbF_3$, $SbF_3Cl_2$ | $Hg_2F_2$, $HgF_2$ | AgF | KF |
| —Si—Cl | ** | ** | — | — | *[c] |
| =P—Cl | ** | ** | — | — | *[c,d] |
| $—SO_2Cl$ | — | * | — | — | **[e] |
| —CO—Cl | * | * | — | ** | **[f] |
| —CCl—CO— | * | * | — | * | ** |
| —CCl—CN | * | * | — | — | ** |
| >C=C—C—Cl | † | † | — | * | **[g] |
| $C_6H_5$—C—Cl | — | — | * | — | * |
| $—C—OCCl_3$ | * | ** | * | — | — |
| $—C—SCCl_3$ | * | ** | * | — | — |
| —C—Cl | † | † | ** | ** | ** |
| >C=C—Cl | † | * | — | — | **[g] |
| $>C=C—CCl_3$ | **[b] | ** | — | — | * |
| $C_6H_5—CCl_3$ | **[b] | ** | — | — | — |
| $>CCl_2,—CCl_3$ | **[b] | ** | ** | — | *[h] |

[a] (*) Applicable; (**) generally used; (†) not feasible.
[b] Usually catalyzed by $SbCl_5$.
[c] Also $KSO_2F$.
[d] Also NaF.
[e] Also $KHF_2$.
[f] Also TlF.
[g] Also RbF, CsF.
[h] Also $Na_2SiF_6$ for partial replacement.

### Replacement by Means of Hydrogen Fluoride

Except for some substitutions of fluorine for halogens bound to silicon which were accomplished using aqueous hydrofluoric acid, anhydrous hydrogen fluoride is usually necessary for carrying out conversions of halogen derivatives to fluorine compounds. This reagent does not, as a rule, replace single halogen atoms bound to carbon **[95]**. It is especially suited for partial or complete replacement of halogens by fluorine in all kinds of organic geminal polyhalides. Such replacements are very easy in polyhalogeno groups adjacent to double bonds or aromatic nuclei **[96–98]**. In all other instances, very energetic conditions such as temperatures in excess of 100°C and the corresponding pressures must be applied **[95]** unless catalysts are used. The classic *antimony trichloride–antimony pentachloride* catalyst suitable for liquid-phase fluorinations is still in use **[99,100]**. However, vapor-phase catalytic processes using activated charcoal, especially impregnated with ferric chloride **[101]**, or special catalysts prepared from chromium, thorium, and other compounds on alumina, take over **[102]**.

(40) $C_6H_5CCl_3 \rightarrow C_6H_5CF_3$

- **[96]** HF, 0°C: 75–95%
- **[97]** HF, 40°C, 1.5 atm: 70–75%
- **[98]** HF, 135–145°C, 15 atm: 89%

(41) $CCl_3CH_2CCl(CH_3)_2 \xrightarrow[130°C]{HF} CF_3CH_2CCl(CH_3)_2$ 40% **[95]**

(42) $CCl_4 \rightarrow CCl_3F + CCl_2F_2 + CClF_3$

- **[99]** HF, $SbCl_3$, $Cl_2$, 110°C, 30 atm: 9%, 90%, 0.5%
- **[101]** HF, $FeCl_3$, C, 300°C: 20%, 75%

(43) $CCl_3CN \xrightarrow[450°C,\ 5\ hr]{HF,\ CrO_3F_2} CCl_2FCN + CClF_2CN + CF_3CN$ **[100]**

32.4% 51% 3.5%

### Replacement by Means of Antimony Fluorides

Antimony fluorides, especially antimony trifluoride, are reagents of choice for partial or total replacement of halogen atoms in geminal polyhalides by fluorine. Since metal apparatus is not always necessary and the work under atmospheric pressure can be carried out in glass or plastics, antimony trifluoride is preferred to anhydrous hydrogen fluoride in small-scale preparations in conventional laboratories. Its action on organic halogen derivatives resembles that of anhydrous hydrogen fluoride, and the reactivity of the reagent can be increased by converting part or all of the *antimony to the pentavalent state* by adding to it varying amounts of chlorine, bromine, or antimony pentachloride. The reactivity and potency of antimony fluorides increases in the series

(44) $\underline{SbF_3} < SbF_3 + SbCl_3 < \underline{SbF_3 + SbCl_5} < SbF_3Cl_2 < SbF_5$

Single, isolated halogen atoms are usually not replaced by fluorine using antimony trifluoride **[103]**. In the reaction of antimony trifluoride with geminal polyhalides, the number of halogen atoms replaced by fluorine depends on the amount of the antimony reagent, which should be used in a slight excess above the stoichiometric ratio. For partial replacement, milder conditions (less active agent, lower temperature, shorter reaction time) should be applied **[104–107]**. For total replacement, the presence of pentavalent antimony is preferable, although not always necessary **[103,107,108]**. Halogen atoms in methylene or methyl groups adjacent to double bonds **[109–111]** or aromatic nuclei **[112]** are replaced by fluorine especially readily. A few typical reactions exemplify the experimental conditions:

(45) $CCl_3CH_2CH_2Cl \xrightarrow[20°C]{SbF_3,\ SbCl_5} CF_3CH_2CH_2Cl$ 75% **[105]**

(46) $CCl_3CCl_2OCH_3 \xrightarrow[\text{reflux 2 hr}]{SbF_3} CCl_3CF_2OCH_3$ 85.5% **[108]**

(47) $CH_3SCCl_3 \xrightarrow[95°C,\ 1.25\ hr]{SbF_3,\ SbCl_5} CH_3SCF_3$ 73%

$CH_3SCCl_3 \xrightarrow[140°C,\ 15\ min]{SbF_3} CH_3SCClF_2 + CH_3SCF_3$ 32.4% 37.8% **[107]**

(48) $CCl_2{=}CCl{-}CCl{=}CCl_2 \xrightarrow[135°C,\ 155°C,\ 6\ atm,\ 2\ hr]{SbF_3Cl_2} CF_3CCl{=}CClCF_3$ 73% **[109]**

$CCl_2{=}CCl{-}CCl{=}CCl_2 \xrightarrow[110\text{–}140°C,\ 1.75\ hr]{SbF_3,\ SbCl_5} CF_3CCl{=}CClCF_3$ 60% **[110]**

(49) cyclo-($CCl{=}CCl{-}CCl_2{-}CCl_2{-}CCl_2$) $\xrightarrow[\text{7 atm, 1 hr, distn.}]{SbF_3,\ SbF_3Cl_2}$ cyclo-($CCl{=}CCl{-}CF_2{-}CF_2{-}CF_2$) 72.5% **[111]**

(50) $Cl{-}C_6H_4{-}CCl_2{-}C_6H_4{-}Cl \xrightarrow[135°C,\ 5\ min]{SbF_3,\ Br_2} Cl{-}C_6H_4{-}CF_2{-}C_6H_4{-}Cl$ 52.5% **[112]**

(51)

$$CCl_3CH_2CH_2Cl \xrightarrow[165^\circ C\ 2\ hr]{SbF_3,\ SbCl_5} 75\%\ CF_3CH_2CH_2Cl + CClF_2CH_2CH_2Cl + CCl_2FCH_2CH_2Cl$$

[105]; [104] $SbF_3$, $SbF_3Cl_2$, $C_6H_5CF_3$, 20°C→reflux: 61% $CF_3CH_2CH_2Cl$ + 10% $CClF_2CH_2CH_2Cl$; [106] $SbF_3$, $SbCl_5$(10%), 120°C: 51% $CClF_2CH_2CH_2Cl$ + 10% $CCl_2FCH_2CH_2Cl$

## Replacement by Means of Silver Fluoride

This, one of the first reagents for introducing fluorine into organic compounds, is now almost outdated. The main disadvantage is the necessity of using at least two moles of the reagent per one halogen atom exchanged, since the reaction product, silver halide, forms an addition compound with one mole of silver fluoride and immobilizes the reagent from the reaction. Nevertheless, it is still occasionally used because it requires very mild conditions. It is especially suited for the replacement of $\alpha$-halogens in ketones, particularly in steroid chemistry **[113,114]**.

(52) steroid-$COCH_2I$ $\xrightarrow[30\text{–}40^\circ C]{AgF,\ CH_3CN,\ H_2O}$ steroid-$COCH_2F$ 63% **[113]**

(53) 1-bromoadamantane (Br) $\xrightarrow[\text{reflux 24 hr}]{AgF,\ C_6H_{12}}$ 1-fluoroadamantane (F) 56% **[114]**

## Replacement by Means of Mercury Fluorides

Both mercurous and mercuric fluorides are used for substituting fluorine for halogens. Although capable of replacing many kinds of halogens, they are not destructive to sensitive organic compounds whose functional groups could be attacked by strong fluorinating reagents. Mercuric fluoride is more reactive than mercurous fluoride. However, if the latter is mixed with a fraction or an equivalent amount of bromine or iodine, such a mixture can accomplish the same type of reaction as pure mercuric fluoride **[115]**. Both reagents are suitable for replacing single halogen atoms as well as for partial or total exchange of halogens for fluorine in geminal di- and tri-halogen derivatives. Halogen atoms in geminal polyhalogen derivatives are replaced preferentially to single halogens **[116–118]**.

(54) $$CH_2BrCHBrCO_2CH_3 \xrightarrow[140\text{–}150°C,\ 2.5\ hr]{HgF,\ I_2} CH_2FCHBrCO_2CH_3 \quad 29\%$$ [115]

(55) $$CH_2BrCBr_2CO_2CH_3 \xrightarrow[\text{distn, } in\ vacuo]{HgF_2} CH_2BrCBrFCO_2CH_3 \quad 24\%$$ [116]

(56) $$CCl_3SCl \xrightarrow[CH_2Cl_2,\ \text{reflux 3 hr}]{HgF_2} CCl_2FSCl \quad 51.5\%,\ 39\%$$ [117] [118]

## Replacement by Means of Potassium Fluoride

Originally, potassium fluoride was believed to be capable only of replacing activated halogen atoms by fluorine. Gradually, it developed into one of the most universal of fluorinating agents with the potentiality of converting almost any type of halogenated compound to a fluoro derivative. Except for a few cases where even concentrated aqueous solution can be used (for instance, in converting sulfonyl chlorides to sulfonyl fluorides), anhydrous potassium fluoride is usually required for metathetical reactions. Drying is carried out for several hours in an oven at atmospheric pressure at 120–150°C, or in a vacuum at 100°C. The reagent is used in a large excess and in a finely powdered form. Anhydrous, strongly polar solvents, especially of glycol type, make for good yields. A molecular compound of potassium fluoride and sulfur dioxide formulated as potassium fluorosulfinate $KSO_2F$ was found in some cases superior to potassium fluoride alone **[119]**.

Experimental conditions used in the replacement reactions vary according to the reactivity of individual halogen atoms to be exchanged. Very easy replacement occurs in acyl halides **[120,121]** and halogen compounds having halogen atoms in $\alpha$-positions to carbonyl or carboxyl groups **[122,123]**, or a double bond **[124]**.

(57) $$CH_3COCl \xrightarrow[100°C]{KF,\ CH_3CO_2H} CH_3COF \quad 76\%$$ [120]

(58) $$ClCO_2C_2H_5 \rightarrow FCO_2C_2H_5$$
[121] KF, $CH_3COCH_2COCH_3$, 50–60°C, $h\nu$ — 51%
[131] TlF, 20°C, 15 hr, 100°C, 30 min — 58%

(59) $$ClCH_2CO_2C_2H_5 \xrightarrow[110\text{–}140°C]{KF,\ CH_3CONH_2} FCH_2CO_2C_2H_5 \quad 60\text{–}63\%$$ [123]

(60) $$CF_2{=}CFCCl_2F \xrightarrow[47°C,\ 17\ hr]{KF,\ HCON(CH_3)_2} CF_3CF{=}CClF \quad 62\%$$ [124]

A nonactivated halogen in aliphatic chains requires rather forcing conditions **[125]**. The same is true of aromatic halogens unless they are activated for a nucleophilic reaction by proper substituents such as a carboxylic group or a nitro group in *ortho* and/or *para* positions **[126]**. The replacement of halogen atoms in aromatic polyhalogen derivatives leading to halogenofluoro or perfluoro compounds calls for extremely drastic reaction conditions **[127,128]**.

(61) $$C_6H_{13}Cl \xrightarrow[175\text{–}185°C]{KF,\ (CH_2OH)_2,\ (HOCH_2CH_2)_2O} C_6H_{13}F \quad 54\%$$ **[125]**

(62) 1-Cl-2,4-dinitrobenzene → 1-F-2,4-dinitrobenzene **[126]**

KF, $C_6H_5NO_2$, 195–210°C: 76%
KF, $HCON(CH_3)_2$, 95–100°C: 77%
KF, $CH_3SOCH_3$, 95–100°C: 78%
KF, $CH_2(CN)_2$, 125–140°C: 81%

(63) $$C_6Cl_6 \xrightarrow[450\text{–}500°C]{KF} C_6F_6 + C_6ClF_5 + C_6Cl_2F_4 + C_6Cl_3F_3$$ 21%, 20%, 14%, 12% **[127]**

(64) Pentachloropyridine
KF, $(CH_2)_4SO_2$, 190–210°C, 36 hr → 3,5-dichloro-2,4,6-trifluoropyridine 68% + 3-chloro-2,4,5,6-tetrafluoropyridine 7%
KF, 480°C, 24 hr → 3-chloro-2,4,5,6-tetrafluoropyridine 7.7% + pentafluoropyridine 68.5% **[128]**

## Replacement by Means of Other Fluorides

Of the other alkali fluorides, *sodium fluoride* is less reactive, *rubidium* and *cesium fluorides* more reactive than potassium fluoride. The former is used for preparing fluorophosphates from chlorophosphates **[129]**, the latter two for the replacement of aromatic halogens by fluorine (Table 13) **[130]**.

(65) $$[(CH_3)_2CHO]_2P(O)Cl \xrightarrow[\text{reflux}]{NaF,\ CCl_4} [(CH_3)_2CHO]_2P(O)F \quad 60\text{–}70\%$$ **[129]**

*Thallous fluoride* was successfully used for converting chloroformates to fluoroformates **[131]** (see pp. 36,40). *Arsenic fluoride* is very promising because of easy handling **[132]**. *Sodium fluorosilicate* is capable of partial replacement of chlorine in polychloro derivatives and is attractive from the industrial point of view since it makes use of fluorine in waste gases from the large-scale production of superphosphates **[133]**.

**Table 13. Efficiency of Alkali Metal Fluorides in Replacement of Halogens by Fluorine [130]**

| Chloro-compound | | Yield of the corresponding fluoro-compound, % | | | | |
|---|---|---|---|---|---|---|
| | | LiF | NaF | KF | RbF | CsF |
| 2-nitrochlorobenzene ($NO_2$, Cl) | (a) | — | — | — | 6 | 80 |
| | (b) | — | — | 6 | 16 | 45 |
| 2,4-dinitrochlorobenzene ($O_2N$, $NO_2$, Cl) | (a) | — | — | 51 | 88 | 98 |
| | (b) | — | — | 7 | 39 | 73 |
| 2,4,6-trinitrochlorobenzene ($O_2N$, $NO_2$, $NO_2$, Cl) | (a) | 4 | 17 | 92 | — | — |

[a] No solvent, 180–200°C.
[b] Dimethylformamide, 95–155°C.

(66) $CCl_3CCl_3$

[132] $AsF_3$, $SbCl_5$, reflux 4 hr → $CCl_3CCl_2F$ (35%) + $CCl_2FCCl_2F$ (5%)

[133] $Na_2SiF_6$, 280°C, 35 atm, 2 hr → $CCl_3CCl_2F$ (22%) + $CCl_2FCCl_2F$ (31%) + $CCl_2FCClF_2$ (3%)

## REPLACEMENT OF OXYGEN BY FLUORINE

One of the most convenient methods of preparation of halogen derivatives by replacement of oxygen by halogens was long considered unfeasible for fluorine compounds. Only recently have good preparative methods been worked out for obtaining fluorine compounds from epoxides, sulfonic esters, alcohols, aldehydes, ketones, and acids. Some special reagents were developed for these purposes.

### Cleavage of Ethers and Epoxides

Ethers are stable toward hydrogen fluoride and similarly acting fluorinating agents. However, one exception is worth quoting: Conversion of 1-methoxybicyclo(2,2,2)octane to 1-fluorobicyclo(2,2,2)octane by acetyl fluoride in the presence of boron trifluoride **[134]** [Equation (78)].

On the other hand, the carbon–oxygen bond in epoxides (oxiranes) is readily split by anhydrous hydrogen fluoride or its salts with amines to give vicinal fluorohydroxy compounds **[135]**. The same reaction is achieved by means of boron trifluoride etherate and was successfully applied especially in steroids **[136]**.

(67)

$$CH_3COO\text{-steroid epoxide} \xrightarrow[20°C,\ 30\ min]{BF_3\cdot(C_2H_5)_2O;\ C_6H_6} CH_3COO\text{-steroid (F, OH)}$$ [136]

## Cleavage of Sulfonic Esters

Alkyl esters of sulfuric acid, methanesulfonic acid, benzenesulfonic acid, and *p*-toluenesulfonic acid react with potassium fluoride in such a way that the sulfonyloxy group is nucleophilically displaced by fluoride ion **[137,138]**. The reaction may be combined with the preparation of the sulfonic ester by using alcohol and methasulfonyl fluoride and heating this mixture with potassium fluoride **[139]**.

(68) $$p\text{-}CH_3C_6H_4SO_2OR \xrightarrow[250°C,\ 500\ min]{KF} RF$$ **[138]**

| R= | $CH_3$ | $C_2H_5$ | $C_3H_7$ | $(CH_3)_2CH$ |
|---|---|---|---|---|
| % | 89.4 | 85.6 | 63.3 | 50.7 |

(69) $$ClCH_2CH_2OH \xrightarrow[2.\ KF,\ (OHCH_2CH_2)_2O,\ 100°C,\ 300\ min]{1.\ CH_3SO_2F,\ KF,\ 20°C,\ 24\ hr} ClCH_2CH_2F\quad 53\%$$ **[139]**

## Replacement of Hydroxylic Group by Fluorine

Contrary to traditional views, quite a few examples of direct replacement of a hydroxylic group by fluorine can be found in the literature. Anhydrous **[140]** and even aqueous hydrogen fluoride **[141]** can be used for converting *alcohols to fluorides*. Only tertiary fluorides were successfully prepared in this way.

(70) $$(CH_3)_3COH \xrightarrow[60°C]{HF(60\%)} (CH_3)_3CF\quad 60\%$$ **[141]**

Of more general applicability is the reaction of hydroxyl-containing compounds with reagents which yield unstable intermediates decomposing to organic fluorides.

(71) $$R\text{—}OH + F\text{—}YX \xrightarrow{-HX} R\text{—}O\text{—}Y\text{—}F \longrightarrow R\text{—}F + O{=}Y$$

X=F, Cl; Y=>CO, >CS, >SO, >SF₂, >SOF₂, CHClFC(NR₂)<

Out of many possibilities, the most useful reactions will be exemplified in the following equations. Alcohols are first converted to alkyl chloroformates by phosgene, the chloroformates are converted to *fluoroformates*, and these decomposed with the evolution of carbon dioxide **[142]**. The fluoroformates can be prepared directly from alcohols or phenols using carbonyl difluoride or carbonyl chlorofluoride **[143]**.

(72) $ROH \xrightarrow{COCl_2} ROCOCl \xrightarrow[\text{or TlF}]{KF} ROCOF$; $ROCOF \xrightarrow[100\text{–}110°C]{C_5H_5N} RF$ 58–75%; $ROCOF \xrightarrow[0\text{–}50°C]{BF_3 \cdot (C_2H_5)_2O} RF$ 36–86% **[142]**

(73) 3-hydroxyphenol (OH, OH) $\xrightarrow[(C_4H_9)_3N,\ C_7H_8]{COClF,\ 60°C,\ 18\ atm}$ $C_6H_4(OCOF)_2$ (OCOF, OCOF) 94.5% $\xrightarrow[5\text{–}6\ sec]{700°C}$ $C_6H_4F(OCOF)$ (F, OCOF) 38% + $C_6H_4F_2$ (F, F) 7.3% **[143]**

Another method is based on the conversion of an alcohol to a chlorosulfinyl ester, which, on treatment with potassium fluorosulfinate, gives a *fluorosulfinyl ester*, and this is decomposed by heat to an alkyl fluoride **[144]**:

(74) $C_4H_9OSOCl \xrightarrow[100°C,\ 2\ hr,]{KSO_2F} C_4H_9OSOF$; $C_4H_9OSOF \xrightarrow[120°C,\ 2\ hr,\ 200°C,\ 15\ min]{(C_4H_9)_3N} C_4H_9F$ 53.2%; $C_4H_9OSOF \xrightarrow[200°C,\ 1\ hr,\ autoclave]{C_6H_5N(CH_3)_2} C_4H_9F$ 88.6% **[144]**

Good results were reached in the reaction of *alcohols with sulfur tetrafluoride* **[145]**, *triphenyldifluorophosphorane* **[146]**, and especially *diphenyltrifluorophosphorane* **[147]**.

(75) $$C_6F_5CH(OH)CF_3 \xrightarrow{SF_4,\ C_5H_{10}} C_6F_5CHFCF_3\quad 90\%$$ **[145]**

(76) $$C_7H_{15}CH_2OH \xrightarrow[150°C,\ 7\ hr]{(C_6H_5)_3PF_2,\ CH_3CN} C_7H_{15}CH_2F\quad 68\%$$ **[146]**

Of a great preparative value is the reaction of *alcohols with 2-chloro-1,1,2-trifluorotriethylamine*, which is easily prepared by the addition of diethylamine to chlorotrifluoroethylene (p.134). The conversion of alcohols to fluorides takes place at very mild conditions and gives fair yields. The method was successfully applied to primary **[148]**, secondary **[149]**, and even some tertiary alcohols **[150]**, although side reactions such as dehydration and rearrangement may sometimes decrease the yields **[151]**. With nonsymmetrical alcohols, inversion of configuration as well as retention were observed **[151]**.

(77) $C_4H_9OH + CHClFCF_2N(C_2H_5)_2 \xrightarrow{\text{spont.}} C_4H_9F + CHClFCON(C_2H_5)_5$ **[148]**
66.5% 72.5%

(78) OH (bicyclic alcohol) $\xrightarrow[60–65°C]{CHClFCF_2N(C_2H_5)_2}$ F (bicyclic fluoride) $\xleftarrow[10–20°C, 3 hr]{CH_3COF, BF_3 \cdot H_3PO_4}$ $OCH_3$ (bicyclic ether)
63.5% [150] 70% [134]

## Replacement of Carbonyl Oxygen by Fluorine

Replacement of carbonyl oxygen by fluorine converts aldehydes and ketones to geminal difluorides. The reaction is accomplished by means of *sulfur tetrafluoride*, alone, or combined with catalysts such as boron trifluoride, hydrogen fluoride, or even small amounts of water, which hydrolyzes a part of sulfur tetrafluoride to hydrogen fluoride **[152,153]**. The catalysts decrease the reaction temperature.

(79) $OCH—C_6H_4—CHO \xrightarrow[150°C, 90 atm, 13 hr]{SF_4} F_2CH—C_6H_4—CHF_2$ 80% **[152]**

(80) $OCOCH_3$ (steroid ketone) $\xrightarrow[20–25°C]{SF_4, 20\% H_2O, CH_2Cl_2}$ $OCOCH_3$ (steroid gem-difluoride, F, F) 82% **[153]**

The same reaction can be achieved by *phenylsulfur trifluoride* **[154]**, or better still, by *selenium tetrafluoride*, which reacts even at room temperature **[155]**.

(81) $C_6H_5COC_6H_5 \xrightarrow[150–160°C, 2 hr, 185–205°C, 3 hr]{C_6H_5SF_3, TiCl_4} C_6H_5CF_2C_6H_5$ 45% (55% recovered) **[154]**

(82) $C_6H_5COCOC_6H_5 \xrightarrow[24°C, 2 hr]{SeF_4} C_6H_5CF_2CF_2C_6H_5$ 80% **[155]**

## Conversion of Carboxylic Group to Trifluoromethyl Group

*Sulfur tetrafluoride* is a unique reagent for the conversion of a carboxylic group to a trifluoromethyl group **[156]**. The first step in this reaction is the replacement of the hydroxylic group by fluorine. The intermediate acyl fluoride can be intercepted when smaller amounts of sulfur tetrafluoride are used.

(83)

$$HOCO(CH_2)_8COOH \xrightarrow[120°C,\ 6\ hr]{3\ SF_4} FCO(CH_2)_8COF\ (21\%) + CF_3(CH_2)_8COF\ (45\%) + CF_3(CH_2)_8CF_3\ (27\%)$$

$$HOCO(CH_2)_8COOH \xrightarrow[120°C,\ 6\ hr]{6\ SF_4} CF_3(CH_2)_8CF_3\ \ 87\%$$

[156]

Not only free carboxylic acids, but also their functional derivatives such as esters, anhydrides, and amides were successfully transformed to trifluoromethyl compounds. The relative reactivity of functional groups toward sulfur tetrafluoride decreases in the sequence **[156]**

(84)

$$C{-}OH > CO > \begin{matrix} COOH \\ CONH_2 \end{matrix} > \begin{matrix} COOCO \\ COOR \end{matrix} > C{-}Hal$$

(85)

$$C_6H_5CO_2CH_3 \xrightarrow[300°C,\ 6\ hr]{SF_4} C_6H_5CF_3\ (55\%) \qquad C_6H_5CONH_2 \xrightarrow[150°C,\ 8\ hr]{SF_4} C_6H_5CF_3\ (13\%)$$

[156]

## REPLACEMENT OF NITROGEN BY FLUORINE

Replacement of nitrogen by fluorine in the aliphatic series is very rare. Diazoacetone and anhydrous hydrogen fluoride give fluoroacetone **[157]**; ethyl diazoacetate and hydrogen fluoride with N-bromosuccinimide give ethyl bromofluoroacetate **[158]**.

(86)

$$CH_3COCHN_2 \xrightarrow{HF,\ (C_2H_5)_2O} CH_3COCH_2F\ \ 74\%$$

[157]

(87)

$$N_2CHCO_2C_2H_5 \xrightarrow[-70°\rightarrow 20°C,\ 10\ hr]{HF,\ \begin{matrix} CH_2{-}CO \\ CH_2{-}CO \end{matrix}\!\!>NBr,\ (C_2H_5)_2O} BrFCHCO_2C_2H_5\ \ 59\%$$

[158]

Exceptionally, an aliphatic primary amine was converted to the corresponding fluoride resulting from the replacement of the diazotized amino group. But the domain of this reaction is in aromatic chemistry, where it is the most common way of introducing fluorine into aromatic nuclei. Diazotization of primary aromatic amines followed by decomposition of the diazonium salts leads to good to excellent yields of aromatic fluorides. The most common variety of this reaction is the *Balz–Schiemann reaction:* primary aromatic amines are converted to aromatic *diazonium fluoroborates* by treatment with fluoroboric acid, or with any mineral acid followed by the addition of fluoroboric acid or sodium or ammonium fluoroborate. The diazonium fluoroborates, generally sparingly soluble, are isolated, dried, and decomposed by heating **[159,160]**. Some unstable diazonium fluoroborates are better decomposed when diluted with inert material such as sand **[160]**, or even *in situ* in the medium in which they have been prepared **[161]**.

(88) $C_6H_5NH_2 \xrightarrow[\text{2. } HBF_4\text{, (40\%)}]{\text{1. } NaNO_2\text{, } HCl} C_6H_5N_2BF_4$ (63%) $\xrightarrow[\text{2. distn.}]{\text{1. heat}} C_6H_5F$ 97% **[159]**

(89) 2-amino-6-fluoropyridine $\xrightarrow[-8 \text{ to } -10°C\text{, 1 hr, 0–50°C, distn.}]{NaNO_2\text{, } HBF_4}$ 2,6-difluoropyridine 27% **[161]**

Also, diazotization of aromatic amines in anhydrous hydrogen fluoride **[162]** and even concentrated aqueous hydrofluoric acid **[163]** followed by heating of the solution of the *diazonium fluorides* until they decompose proved successful.

(90) $H_2N–C_6H_4–CO_2H \xrightarrow[0° \to 100°C]{NaNO_2\text{, HF}} F–C_6H_4–CO_2H$ 98% **[162]**

(91) 4-amino-2-hydroxybenzoic acid $\xrightarrow[\text{2. 40–70°C, 80°C, 1 hr}]{\text{1. } NaNO_2\text{, HF(70\%), 15–20°C}}$ 87.2% 4-fluoro-2-hydroxybenzoic acid **[163]**

4-amino-2-hydroxybenzoic acid $\xrightarrow[5–10°C]{NaNO_2\text{, } HBF_4}$ 2-hydroxy-4-($N_2BF_4$)benzoic acid ($CO_2H$, OH, $N_2BF_4$) 78.4%

→ heat, dry: 21.9%; → xylene, 90°C, reflux 0.5 hr: 32.2% (4-fluoro-2-hydroxybenzoic acid)

Other complex diazonium salts such as *diazonium fluorosilicates* **[164]** and *diazonium fluorophosphates* **[165]** also decompose to give the corresponding aromatic fluoro derivatives, sometimes even in yields superior to those obtained in the Schiemann reaction or diazotization in hydrogen fluoride.

(92) 2-aminopyridine $\xrightarrow[\text{2. 40–50°C, 1 hr}]{\text{1. } NaNO_2\text{, } H_2SiF_6 \text{ (30\%), 5°C}}$ 2-fluoropyridine 42% **[164]**

(93)

$CO_2H$ / $NH_2$ $\xrightarrow[\text{2. } HPF_6]{\text{1. } NaNO_2,\ HCl}$ 79% $CO_2H$ / $N_2PF_6$ $\xrightarrow{\text{heat}}$ 78% $CO_2H$ / F (57%)

$\xrightarrow[0\text{–}100°C]{NaNO_2,\ HF}$ [162]

[165]

**Table 14. Survey of Methods for Introducing Fluorine into Organic Compounds**

| Starting material | Product | | | | |
|---|---|---|---|---|---|
| | Monofluorides | Difluorides | | Polyfluorides | |
| | | Vic. | Gem. | Trigem. | Poly- |
| *Hydrocarbons* | | | | | |
| Paraffins | $F_2$ | — | — | — | $F_2$,HVF[a] |
| Olefins | HF | $F_2$,HF/$PbO_2$ | — | — | — |
| Acetylenes | — | — | HF | — | — |
| Alicyclics | — | — | — | — | $F_2$,HVF |
| Aromatics | $XeF_2$,$ClF_3$ | — | — | — | $F_2$,HVF |
| *Halides* | | | | | |
| Monohalo | AgF,HgF, $HgF_2$,KF | — | — | — | $F_2$,HVF |
| Dihalo vic. | — | AgF,HgF, $HgF_2$,KF | — | — | — |
| Dihalo gem. | — | — | $HgF_2$,KF, HF,$SbF_3$ | — | — |
| Trihalo gem. | — | — | — | $HgF_2$,KF, HF,$SbF_3$ | — |
| Polyhalo | — | — | — | — | KF |
| Allylic | — | — | — | — | — |
| Vinylic | — | — | — | — | — |
| Benzylic | — | — | — | — | — |
| Aromatic | — | — | — | — | — |
| *Functional compounds* | | | | | |
| Alcohols | HF,$SF_4$, $Ph_3PF_2$, CTTA[b] | — | — | — | — |
| Aldehydes, ketones | — | — | $SF_4$, $PhSF_3$ | — | — |
| Acids | — | — | — | $SF_4$, $PhSF_3$ | — |
| Esters | — | — | — | $SF_4$ | — |
| Amides | — | — | — | $SF_4$ | — |
| Sulfo esters | KF | KF | — | — | KF |
| Amines | $NaNO_2$ +HF[c] | — | — | — | $NaNO_2$ +HF[c] |

**Table 14 (Continued)**

| Starting material | Product | | | | | |
|---|---|---|---|---|---|---|
| | Unsaturated fluorides | | | Perfluoro compounds | | |
| | Allyl | Vinyl | Benzyl | Paraffins | Alicyclics | Aromatics |
| *Hydrocarbons* | | | | | | |
| Paraffins | — | — | — | $F_2$,HVF, Electro | — | — |
| Olefins | — | — | — | — | — | — |
| Acetylenes | — | HF | — | — | — | — |
| Alicyclics | — | — | — | — | $F_2$,HVF | — |
| Aromatics | — | — | — | — | $F_2$,HVF, Electro | — |
| *Halides* | | | | | | |
| Monohalo | — | — | — | $F_2$,HVF | $F_2$,HVF | — |
| Dihalo vic. | — | — | — | $F_2$,HVF | $F_2$,HVF | — |
| Dihalo gem. | — | — | — | $F_2$,HVF | $F_2$,HVF | — |
| Trihalo gem. | — | — | — | $F_2$,HVF | $F_2$,HVF | — |
| Polyhalo | — | — | — | $F_2$,HVF | $F_2$,HVF | — |
| Allylic | AgF,KF, HF,$SbF_3$[d] | — | — | — | — | — |
| Vinylic | — | KF[d] | — | — | — | — |
| Benzylic | — | — | $HgF_2$,KF, HF,$SbF_3$[d] | — | — | — |
| Aromatic | — | — | — | — | $F_2$,HVF | KF,CsF |
| *Functional compounds* | | | | | | |
| Alcohols | — | — | — | Electro[e] | — | — |
| Aldehydes, ketones | — | — | — | — | — | — |
| Acids | — | — | — | Electro[e] | — | — |
| Esters | — | — | — | — | — | — |
| Amides | — | — | — | — | — | — |
| Sulfo esters | — | — | — | — | — | — |
| Amines | — | — | — | Electro[f] | — | — |

[a] HVF: High-valency fluorides: $AgF_2$, $CoF_3$, $MnF_3$.
[b] CTTA: 2-Chloro-1,1,2-trifluorotriethylamine.
[c] Only aromatic fluorides, also by Schiemann reaction.
[d] In polyhalides.
[e] Perfluoroacids as main products.
[f] Perfluoro-amines from tertiary amines.

A survey of methods for the introduction of fluorine into organic compounds is given in Table 14, and the preparation of fluorine compounds by reactions of other fluorinated organic compounds is outlined in Table 50.

*Chapter 5*

# Analysis of Organic Fluorine Compounds

The analysis of fluorinated compounds is a necessary concomitant to their preparation and reaction. Classical analytical procedures, which are often lengthy and laborious, are gradually giving way to modern physical methods of analysis. These sometimes do not even require work with pure chemical specimens, and allow analyses to be done in mixtures without the isolation of individual components.

## PHYSICAL METHODS OF ANALYSIS OF FLUORINE COMPOUNDS OR THEIR MIXTURES

Physical analytical methods such as infrared spectrophotometry, mass spectroscopy, and nuclear magnetic resonance spectroscopy require extremely expensive equipment compared to the classical analytical methods. However, once such instruments are available, they pay off in the long run because they enable a chemist to carry out, at an unsurpassed speed, qualitative as well as quantitative determination of fluorine and other elements, without preceding isolation of pure components. All these methods are therefore extremely suitable for analytical control of reactions aimed at the synthesis of fluorinated compounds, or their conversions to other derivatives.

### Infrared Spectroscopy

Fluorine in organic compounds does not have any characteristic absorption that would identify it among other absorption bands. It was found that the carbon–fluorine stretching vibration shows strong absorption in the region of 1100–1300 $cm^{-1}$ (9.1–7.7 microns) **[166,167]**, but this is hardly enough for identification of organic fluorine compounds. It is therefore necessary to know peaks characteristic for the compound to be identified or characterized, and to look for these peaks in the spectrum of the investigated derivative. Spectra of many organic fluorine compounds have been recorded **[168]**. However, in most of the practical examples it is necessary to measure the complete spectrum of the fluorinated derivative first, and use this as a basis for analytical measurements of mixtures of this compound. In an ideal case, a fluorinated compound shows peaks in the region where there is no absorption by the other components of the mixture to be analyzed. This occasionally happens. The example of the infrared spectral analysis of two isomeric bromochlorotrifluoroethanes shows how a fortunate coincidence of the presence of absorption peaks of one isomer in the region where the other isomer does not absorb at all helped carry out qualitative as well as quantitative analysis of an inseparable mixture of compounds **[169]**. Infrared spectroscopy is not limited to solid or liquid compounds or their solution. Since it is very sensitive, it can be applied also to gaseous compounds whose spectra can be recorded at pressures as low a few mm Hg.

In addition to the infrared spectra published in current literature and in the well-known Sadtler Collection, there exist large collections of infrared spectra which can be solicited from several scientific centers, such as the Scientific Documentation Center, Dunfermline, Scotland.

## Mass Spectroscopy

The mass spectroscopy of fluorinated compounds is not yet developed to such an extent as to identify all important cracking patterns. Consequently, authentic samples are usually needed for identifying fragments occurring in mass-spectroscopic analysis. In addition to the determination of molecular weights of unknown compounds, this method is suitable for analysis of multicomponent mixtures, and is very useful in the analysis of mixtures of fluorocarbons. It requires extremely small amounts of material, and gives a great deal of information in a single measurement. An especially

**Table 15. Relative Abundances of the Ions (%) in the Mass Spectra of Perfluoroparaffins [172]**

| Ion | *m/e* | $CF_4$ | $C_2F_6$ | $C_3F_8$ | $C_4F_{10}$ | $C_5F_{12}$ |
|---|---|---|---|---|---|---|
| $C_4F_9^+$ | 219 | — | — | — | 2.6 | — |
| $C_4F_7^+$ | 181 | — | — | — | — | 3.3 |
| $C_3F_7^+$ | 169 | — | — | 24.6 | 2.1 | 9.7 |
| $C_3F_6^+$ | 150 | — | — | — | 2.6 | 1.1 |
| $C_3F_5^+$ | 131 | — | — | — | 8.4 | 6.5 |
| $C_2F_6^+$ | 119 | — | 41.3 | 9.0 | 18.3 | 29.5 |
| $C_2F_4^+$ | 100 | — | — | 6.6 | 8.4 | 7.2 |
| $C_3F_3^+$ | 93 | — | — | — | 1.2 | 2.1 |
| $CF_3^+$ | 69 | 100.0 | 100.0 | 100.0 | 100.0 | 100.0 |
| $CF_2^+$ | 50 | 11.8 | 10.1 | 9.3 | 4.2 | 3.1 |
| $CF^+$ | 31 | 4.9 | 18.3 | 28.8 | 12.2 | 9.2 |

useful apparatus for carrying out mass-spectrometric analysis is the combination of a gas–liquid chromatograph with mass spectrometer.

Information about mass spectroscopy of fluorinated compounds can be found in tables **[170,171]**, and in the review literature **[172]**. Table 15 shows the distribution of fragments produced in the mass spectral analysis of fluorocarbons **[172]**.

## Nuclear Magnetic Resonance Spectroscopy

Whereas there is no characteristic feature of fluorinated derivatives in infrared spectrophotometry, fluorine ($^{19}F$) magnetic resonance is unique in that no other element interferes with its determination by this method. Even the more common proton magnetic resonance spectra allow determination of fluorine in organic compounds.

The characteristic chemical downfield shift of proton by fluorine present at the same carbon atom ranges from 5.5 to 6 ppm ($\tau$) (4–4.5$\delta$). Fluorine also causes the splitting of signals of protons bonded to the adjacent carbon atoms. The coupling constants (40–80cps) can be used for identification of fluorinated derivatives.

By far the most reliable method for the determination of fluorine in a compound or in a mixture of compounds is fluorine $^{19}F$ magnetic resonance, in which the signal is given only by a compound containing fluorine. The equipment for this type of spectral determination is not as readily accessible as regular proton magnetic resonance apparatus and requires some special arrangement, but the determination by this method is unambigous and conclusive.

Data necessary for evaluating NMR spectra of fluorinated compounds are available in the pertinent literature **[173–176]**, especially in reference **[174]**.

Tables 16–18 summarize some data on $^{19}F$ nuclear magnetic resonance spectra of fluoro derivatives of methane, ethane, and benzene, respectively.

**Table 16. $^{19}F$ NMR Chemical Shifts (ppm relative to $CF_3CO_2H_{external}$) and Spin–Spin Coupling Constants (cps) in Fluoromethanes and Fluorochloromethanes**

| Compound | $\delta$ | $J_{H\text{-}F}$ | $J_{^{13}C\text{-}F}$ | Ref. |
|---|---|---|---|---|
| $CH_3F$ | 193.4 | 46.4 | 157.5 | 615 |
| $CH_2F_2$ | 64.9 | 50.2 | 235.0 | 615 |
| $CHF_3$ | 0.1 | 79.7 | 274.3 | 615 |
| $CF_4$ | −15.2 | — | 259.2 | 615 |
| $CF_3Cl$ | −52.0 | — | — | 616 |
| $CF_2Cl_2$ | −75.6 | — | — | 616 |
| $CFCl_3$ | −91.9 | — | — | 616 |

**Table 17.** $^{19}F$ **NMR Chemical Shifts (ppm relative to** $CF_3CO_2H_{external}$**) in Fluoroethanes**

| Compound | $\delta_1$ | $\delta_2$ | Ref. |
|---|---|---|---|
| $CF_3CF_2H$ | 11.2 | 62.9 | 617 |
| $CF_3CFH_2$ | 2.4 | 164.4 | 617 |
| $CF_3CFCl_2$ | 5.9 | −2.1 | 618 |
| $CF_3CFBrCl$ | 4.4 | −2.9 | 618 |
| $CF_3CH_3$ | −14.5 | — | 617 |
| $CF_3CH_2Br$ | −8.5 | — | 617 |
| $CF_3CH_2Cl$ | −5.0 | — | 617 |
| $CF_2ClCF_2Cl$ | −7.4 | −7.4 | 618 |
| $CF_2HCF_2H$ | 61.2 | 61.2 | 617 |
| $CF_2ClCFCl_2$ | −11.4 | −7.4 | 618 |
| $CF_2BrCFHCl$ | −16.4 | 66.1 | 618 |
| $CF_2BrCFHBr$ | −18.9 | 68.1 | 618 |
| $CF_2ClCFHBr$ | −12.9 | 72.1 | 618 |
| $CF_2BrCH_2Cl$ | −25.4 | — | 618 |
| $CF_2ClCH_3$ | −31.8 | — | 617 |
| $CF_2BrCH_3$ | −39.9 | — | 617 |
| $CF_2HCH_3$ | 33.1 | — | 617 |
| $CFH_2CH_3$ | 135.9 | — | 617 |

**Table 18.** $^{19}F$ **NMR Chemical Shifts (ppm relative to** $CFCl_3$**) in Monosubstituted Pentafluorobenzenes,** $C_6F_5X$

| X | $\delta_{ortho}$ | $\delta_{meta}$ | $\delta_{para}$ | Ref. |
|---|---|---|---|---|
| F | 162.3 | 162.3 | 162.3 | 611 |
| Cl | 140.6 | 161.5 | 156.1 | 611 |
| Br | 132.5 | 160.6 | 154.7 | 611 |
| I | 119.2 | 162.1 | 152.5 | 611 |
| H | 138.9 | 162.1 | 153.5 | 611 |
| $NO_2$ | 145.8 | 158.8 | 146.5 | 614 |
| $NH_2$ | 163.6 | 165.7 | 174.1 | 613 |
| $NHCH_3$ | 161.9 | 165.2 | 173.1 | 611 |
| $C_6F_5$ | 138.2 | 160.7 | 150.3 | 612 |
| $CH_3$ | 144.0 | 164.3 | 159.3 | 613 |
| $CF_3$ | 140.0 | 160.6 | 147.9 | 613 |
| $OCH_3$ | 158.5 | 164.9 | 164.6 | 613 |
| OH | 163.7 | 165.1 | 170.5 | 611 |
| CN | 132.5 | 159.2 | 143.5 | 613 |
| $CO_2H$ | 139.3 | 161.9 | 151.6 | 614 |

## SEPARATION OF MIXTURES AND PURIFICATION OF COMPONENTS

Conventional methods are used for the separation of fluorinated compounds from their mixtures. In addition to *crystallization* and *fractionation*, *trap-to-trap distillation* of gaseous and volatile compounds at atmospheric pressure or in a vacuum is frequently employed in a system where there is a sufficient difference in the boiling points of the individual components.

For the separation of compounds with boiling points close to each other, *gas chromatography* or *gas–liquid chromatography* are increasingly used, not only on an analytical, but also on the preparative scale. For this purpose, various substrates have been tested such as kieselguhr, cellite, Chromosorb (activated charcoal), firebricks, Teflon-6, etc. impregnated with silicon grase, Kel *F* polymer, silicon elastomers, fluorocarbons such as $C_{21}F_{44}$, or high-boiling esters such as diethylene glycol disuccinate, dibutyl, dioctyl, dinonyl, or diisodecyl phthalate, etc. A new type of fluorinated compound usually requires experimenting in order to find the optimum conditions for the separation of individual components, especially in the case of isomeric compounds with similar boiling points, volatilities, and retention times. General features of this method of separation and purification of compounds as well as their identification by means of retention times or elution volumes are summarized in several papers and reviews **[177–182]**.

Low boiling points of fluorinated derivatives of some nonvolatile compounds make for an easy separation of mixtures of such derivatives by gas–liquid chromatography. An example is chromatographic separation of trifluoroacetyl derivatives of amino acid esters from complex mixtures of amino acids **[183–185]**.

## QUALITATIVE TESTS FOR FLUORINE

Analytical proof of the presence of fluorine is more difficult than that of the other halogens. The simple and relatively dependable Beilstein test fails with fluorine and its compounds, and wet tests are not so distinct and spectacular as those of chlorine, bromine, or iodine.

### Detection of Elemental Fluorine

It is very rarely that elemental fluorine has to be detected. The most dependable test is ignition of paper or a wood splinter in contact with fluorine of sufficient concentration, such as in the vicinity of a leak in an apparatus holding fluorine. This test is specific for fluorine, whereas the convential potassium–iodide starch paper also detects chlorine.

## Detection of Fluoride Ion

Fluoride is one of the strongest complexing agents. Consequently, the majority of analytical tests are based on a change of color of a colored compound of a heavy metal. Such a test for fluoride ion is the change from red to lilac blue of cerium(III) alizarine complexonate (1,2-dihydroxy-3-anthraquinonylmethylamine-N,N-diacetic acid) **[186]**. Other tests show change from red-violet to light yellow. Zirconium or thorium alizarin-sulfonate lakes are essentially decolorized by fluoride. These reactions are very sensitive, which means a certain disadvantage since it is difficult to estimate whether the decoloration is caused by a large or a small amount of the fluoride ion, and consequently whether fluoride is the main compound in a mixture, or only a contaminant. The following procedures were found convenient for the preparation of the reagents **[187,188]**, of which the first can also be used for the preparation of test papers for fluoride. Such papers are available from Gallard-Schlesinger Chemical Manufacturing Company and other firms.

*Zirconium lake:* Alizarin (0.5 g) is dissolved in 200 ml of warm ethanol, a solution of 1.5 g of zirconium tetrachloride in 75 ml of ethanol is added, the excess of ethanol after sedimentation is decanted to form a volume of 25 ml, and the sediment is shaken with 500 ml of water to form a colloidal solution.

To 2.5 ml of a neutral or slightly acidic solution containing fluoride is added an equal volume of concentrated hydrochloric acid and 0.5 ml of the reagent, and the mixture is stirred and allowed to stand for at least 15 sec. In the presence of 0.3 mg of fluoride, the red color turns yellow immediately, with 0.15 mg after 5 sec., and with 0.03 mg after 15 sec. Chlorates, bromates, and iodates interfere.

*Thorium lake:* Two parts of a 0.05% solution of sodium alizarin-sulfonate are mixed with two parts of a buffer prepared by dissolving 9.45 g of chloroacetic acid and 2 g of sodium hydroxide in water and diluting the volume to 100 ml; four drops of 0.05 *M* thorium nitrate solution and six parts of water are added to form the reagent. The solution is not indefinitely stable and must be renewed from time to time. Addition of a fluoride ion changes the color from pink to light yellow.

## Detection of Fluorine in Organic Compounds

Organically bound fluorine must first be converted to fluoride ion. Such mineralization can be carried out by heating of the sample with con-centrated sulfuric acid, fusing with sodium, or combustion using oxidative catalysts and ignition.

Heating of a fluorine-containing organic material with concentrat

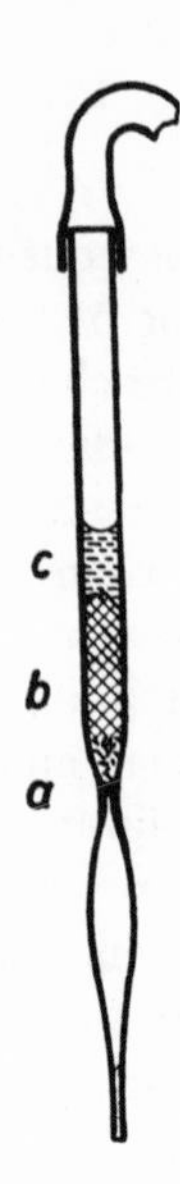


**Fig. 5.** Tube for qualitative test for fluorine in organic compounds: (a) asbestos plug, (b) layer of combustion catalyst (decomposed $AgMnO_4$), (c) layer of distilled water.

sulfuric acid liberates hydrogen fluoride, which is detected by etching of the glass of the test tube, or by turbidity of a hanging drop placed on a glass rod in the mouth of the test tube. The turbidity is caused by silicic acid, produced by the hydrolysis of fluorosilicic acid formed from hydrofluoric acid and glass **[189]**.

Fusing of a 3–5-mg sample with sodium metal in a glass ampoule, immersing of the ampoule still hot in 2 ml of water, and filtering of the solution into 0.1 ml of a buffered alizarinsulfonate–zirconium lake reagent produces change from red to yellow **[190]**.

The best mineralization of organic fluorine is carried out by heating 0.1–1 mg of a fluorine-containing material with 50 mg of a product obtained by decomposing silver permanganate with a free flame in a special tube allowing elution of the mass and filtration through asbestos wool (Fig. 5). Addition of the filtrate to 0.5 ml of a solution of thorium alizarinsulfonate (see above) changes the color from pink to yellow. Nitrogen and sulfur can be detected in the same sample by diphenylamine sulfuric acid and barium nitrate tests, respectively.

## QUANTITATIVE DETERMINATION OF FLUORIDE

The classical gravimetric determination of fluoride ion as calcium uoride or lead chlorofluoride is now outdated.

### olumetric Determination

Of the many suggestions for volumetric determination, the old nanaev–Willard–Winter method **[191]** survived in minor variations. The method is based on decoloration of thorium alizarinsulfonate lake by fluoride ion. If halides, sulfates, phosphates, or arsenates are present in a large excess, separation of the fluoride by "distillation" must precede the determination. The sample is heated with concentrated sulfuric or phosphoric acid in a glass apparatus at a temperature of approximately 135°C. Fluorosilicic acid formed from hydrofluoric acid and glass "distils," the distillate is decomposed with alkalies to fluoride, and after acidification titrated with thorium nitrate using alizarinsulfonate as the indicator.

### Determination Using Fluoride Membrane Electrode

The most universal method for the determination of fluoride is the membrane electrode (Orion Research Inc., Cambridge, Mass.) composed of a monocrystal of lanthanum trifluoride doped with divalent europium **[192–194]**. The monocrystal is placed at the bottom of a tube made of polyvinyl chloride and containing 0.1 *N* sodium chloride and 0.1 *N* sodium fluoride solutions. A silver electrode is immersed in the tube with the above mixture, and the tube itself is immersed in the solution to be analyzed. Potential drop is measured against standard calomel electrode

$$Ag/AgCl,\ Cl(0.1),\ F(0.1)/LaF_3/\text{test solution}/SCE$$

Concentration of fluoride is read from a calibration curve or found by potentiometric titration with calcium, thorium, or lanthanum solutions.

## MINERALIZATION OF FLUORINE IN ORGANIC COMPOUNDS

As in qualitative tests for fluorine in organic fluorine compounds, here too fluorine must be converted to fluoride ion prior to the determination. This conversion is achieved either by decomposition of the organic compound with alkalies or their compounds, or else by combustion of the compound in oxygen atmosphere **[195]**.

### Decomposition with Alkalies

A sample containing 1–5 mg of fluorine is weighed into a small glass vial or capillary tube, inserted into a nickel or stainless steel bomb (Parr bomb) of 2.5 ml capacity, a 30–50-mg piece of sodium is added, and the bomb is sealed and is heated intensively with a Bunsen burner for 5–10 min. After cooling, the bomb is opened, the lid is washed with water in a 100-ml beaker, the content of the bomb is carefully decomposed with a few drops of alcohol or water, dissolved in water, and transferred into the beaker, and the bomb is flushed quantitatively into the beaker. Fluorine is determined by the distillation method followed by the thorium nitrate titration **[196]**.

A very powerful compound for the mineralization of organic fluorine is a complex of diphenylsodium with dimethoxyethane. A sample dissolved in diisopropyl ether is treated with this complex in the cold, cations are removed by means of an anion exchanger—Amberlite IR 120(H)—and the liberated hydrogen fluoride is determined volumetrically **[197]**.

## Combustion in Oxygen

Organic compounds containing fluorine are combusted in a hydrogen–oxygen flame in a quartz apparatus. The combustion gases are absorbed in water or alkalies and the fluoride ion is titrated. The method requires special and rather elaborate and fragile quartz apparatus and a cool-blooded worker **[198,199]**.

Much more simple is the *Schöniger method* of combusting a compound in a quartz flask containing oxygen. Fluorine is converted to hydrogen fluoride, this is absorbed in water, and either titrated with thorium nitrate **[188]** or lanthanum nitrate **[200]**, or else determined spectrophotometrically by measurements of absorption of ferric salicylate–fluoride complex **[201]**.

Two practical modifications of the Schöniger method are listed below as procedures.

A 250-ml quartz flask is charged with 20 ml of distilled water and filled with oxygen. A sample (8–15 mg) wrapped in a piece of ash-free filter paper is inserted into a platinum wire basket fastened to a ground-glass stopper, the paper is ignited, and the whole assembly is placed quickly in the ground-glass neck of the flask. The flask is allowed to stand for 40 min under occasional swirling, after which time the mist formed by burning of the sample is completely absorbed. The ground-glass joint and the walls of the flask are then washed with distilled water; 2 ml of glycine–perchlorate buffer (containing 6.7 g of glycine, 11.0 g of sodium perchlorate, and 11 ml of 1 *N* perchloric acid in 100 ml), 1 ml of 0.05% acqueous solution of sodium alizarinsulfonate, and 0.5 ml of 0.01% aqueous solution of China Blue are added, and the solution is titrated with 0.01 *M* thorium nitrate (5.521 g of thorium nitrate tetrahydrate in 1000 ml of water). The originally green color changes to gray-violet hue. A blank run with distilled water and the reagents usually requires 0.03 ml of the thorium nitrate solution **[188]**.

A sample containing approximately 1.5 mg of fluorine is ignited in the manner described above in the oxygen combustion flask containing 4 ml of distilled water. After 30 min, the stopper and the platinum wire are washed with 2–3 ml of water, the solution is acidified with one drop of 2 *N* perchloric acid, and boiled briefly to expell any carbon dioxide. After the addition of 1 g of hexamethylene tetramine and one drop of 0.5% aqueous solution of haematoxylin, the solution is cooled to room temperature, diluted with 20 ml of acetone or ethanol, and titrated with 0.005 *M* lanthanum nitrate to the color change from yellow to violet. The lantanum nitrate solution containing 4.33 g of lanthanum trinitrate in 2000 ml of water is standardized by means of sodium fluoride. In sulfur-containing samples, the sulfate ions must be removed with barium perchlorate prior to the titration **[200]**.

## Determination of Other Elements in Organic Fluorine Compounds

The composition of an organic fluorine compound is determined by classical methods using catalytic combustion in oxygen. As a rule, fluorine is usually determined separately. Methods for simultaneous determination of *carbon, hydrogen, and fluorine* **[202,203]**, and possibly also other elements, have been worked out, but are of little practical importance. One of them, however, is worth quoting, since it is suitable for the analysis of fluorocarbons. The compound is combusted in oxyhydrogen flame, hydrogen fluoride formed is absorbed in water, and fluoride is titrated with thorium nitrate. Carbon dioxide is expelled from the aqueous solution, is absorbed in an alkaline solution of barium chloride, liberated, and determined gravimetrically by means of ascarite. Nitrogen, sulfur, and other halogens do not interfere **[204]**.

A quick method for the determination of *carbon and hydrogen* is a modification of Pregl elementary analysis and makes use of a special combustion catalyst obtained by decomposition of silver permanganate. Minium placed in the combustion tube is used for absorption of fluorine or hydrogen fluoride. Carbon and hydrogen are determined gravimetrically as carbon dioxide and water, respectively. The same combustion procedure can be used for the separate determination of *nitrogen* by the Dumas method **[205]**.

A quartz combustion tube 10 mm in diameter and 35 cm long is filled as follows: a layer of asbestos, and 8-cm-long layer of the decomposition product of silver permanganate, a plug of asbestos, a 5-cm-long layer of minium ($Pb_3O_4$) on pumice (for absorption of fluorine), and 2-cm-long plug of silver cotton (for absorption of halogens). The combustion is carried out by means of a movable electric furnace at 550°C, water and carbon dioxide are absorbed in Anhydrone and Ascarite, respectively. Combustion time for 15–20-mg samples is 10–15 min, purging of the apparatus another 20 min. For microscale determinations, the tube is only 30 cm long, the combustion catalyst layers are cut down to 5 cm, the minium layer to 3.5 cm, and the combustion takes only 5–7 min with samples of 3–4 mg.

## SPECIAL ANALYSES

Special analyses are often required for specific purposes. As examples, the determination of *water in anhydrous hydrogen fluoride* and in refrigerants will be described. Karl Fischer titration is used in both cases.

Elaborate equipment is needed for taking samples of anhydrous hydrogen fluoride. The whole procedure is explicitly described in the literature **[206]**, and is of much more use for a chemist-producer than a chemist-consumer.

The old method for the determination of *water in refrigerant gases* by passing a 300-g sample through towers packed with asbestos and phosphorus pentoxide at a rate of 30 g/hr was precise to 30 ppm with an accuracy of 0.0001% **[207]**. This determination seems to have given way to a more simple analysis of a solution using the *Karl Fischer* method. A sample of 100 g of refrigerant dissolved in 25 ml of cold methanol is treated with an excess of Karl Fischer reagent, and the water content is determined by back-titration with standard solution of water in methanol **[208]**.

*Chapter 6*

# Properties of Organic Fluorine Compounds

Apart from chemical properties, which will be dealt with in a special chapter, the following features of organic fluorine compounds will be discussed: physical properties, physicochemical properties, and biological properties. Where possible, comparison with the properties of the nonfluorinated parent compounds will be made.

In order to cut down the number of references, citations of original literature have been omitted in cases where the necessary data are summarized in general reference books or in the monographs listed in Table 1.

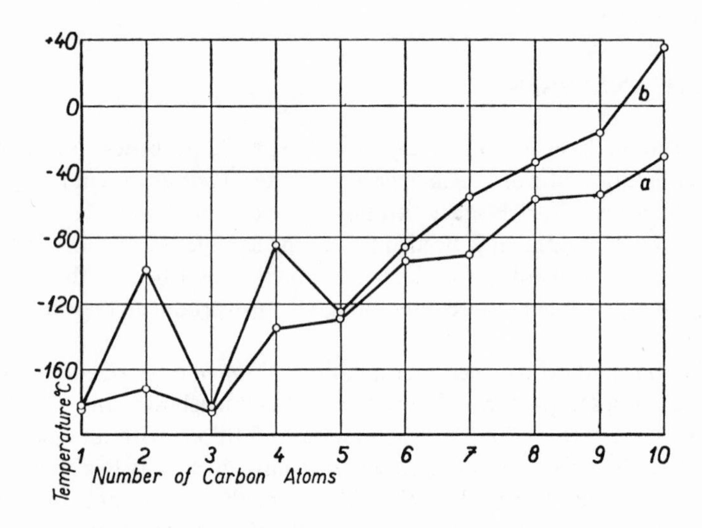


**Fig. 6.** Melting points of paraffins and perfluoroparaffins: (a) paraffins, (b) fluorocarbons.

## PHYSICAL PROPERTIES

### Melting Points

The melting points of perfluoroparaffins show an increase with increasing number of carbon atoms similar to that of the melting points of paraffins. With the exception of carbon tetrafluoride, all the perfluoroparaffins melt higher than the corresponding paraffins. The largest differences show up in $C_2$ and $C_4$ compounds, the smallest in $C_1$, $C_3$, and $C_5$ compounds. Starting with $C_5$, a steady increase, rather than an alternation such as in the series of paraffins, can be observed (Fig. 6).

### Boiling Points

Much better regularity is shown in the boiling points of fluorinated compounds and especially of perfluoroparaffins. The first three perfluoroparaffins show higher boiling points than methane through propane. The boiling point of perfluorobutane is almost the same as for butane, and starting with $C_5$, the boiling points of perfluoroparaffins are consistently lower than those of paraffins. The high volatility of higher perfluoroparaffins

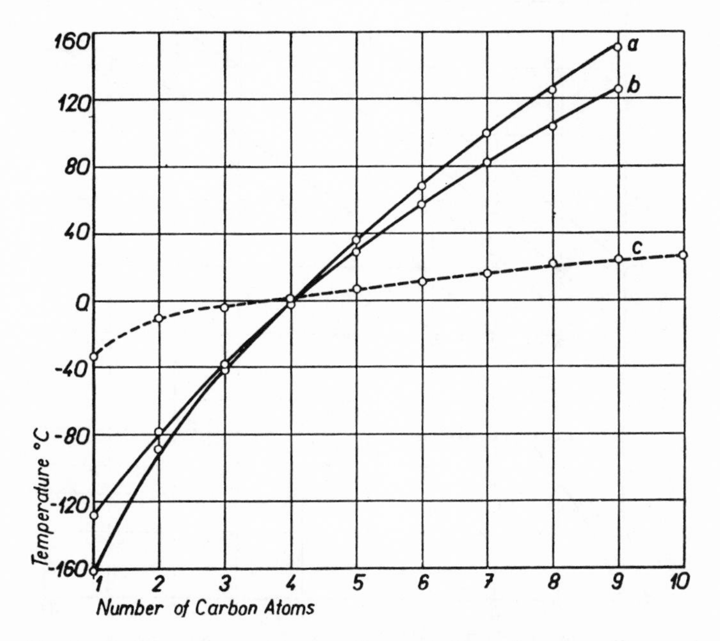


**Fig. 7.** Boiling points of (a) paraffins and (b) perfluoroparaffins, and (c) the differences, boiling point of paraffin minus boiling point of perfluoroparaffin.

is rather surprising considering the high molecular weights of these compounds and polarity of carbon–fluorine bonds (Fig. 7).

A system of parallel lines results from plots of the boiling points of alkyl fluorides and other alkyl halides versus the number of carbon atoms. A similar picture is obtained by plotting the boiling points of perfluoroalkyl halides and perfluoroalkyl hydrides (1H-perfluoroparaffins) (Figs. 8, 9).

Perfluoroalkyl hydrides show higher boiling points than completely fluorinated paraffins, evidently owing to carbon–fluorine–hydrogen bonds. With fluorinated methane and ethane derivatives, the maximum boiling points are exhibited by compounds having approximately equal numbers of hydrogen and fluorine atoms in their molecules. These compounds possess the maximum number of hydrogen bonds (Figs. 10, 11).

In fluorohaloethanes, the more asymmetrical isomers melt consistently higher, and boil consistently lower, than their more symmetrical counterparts (Table 19).

Accurate measurements of *vapor pressures* of many fluorinated refriger-

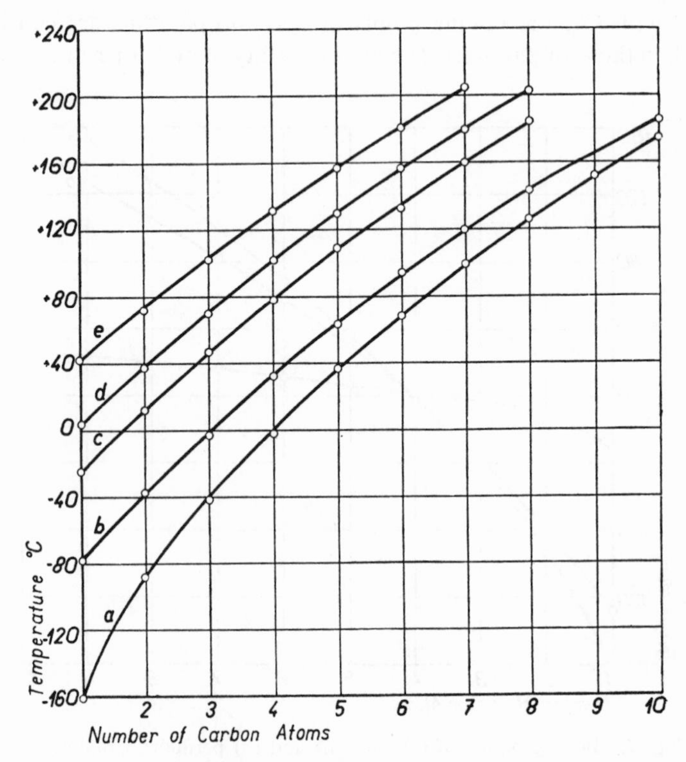


**Fig. 8.** Boiling points of (a) paraffins, (b) alkyl fluorides, (c) alkyl chlorides, (d) alkyl bromides, and (e) alkyl iodides.

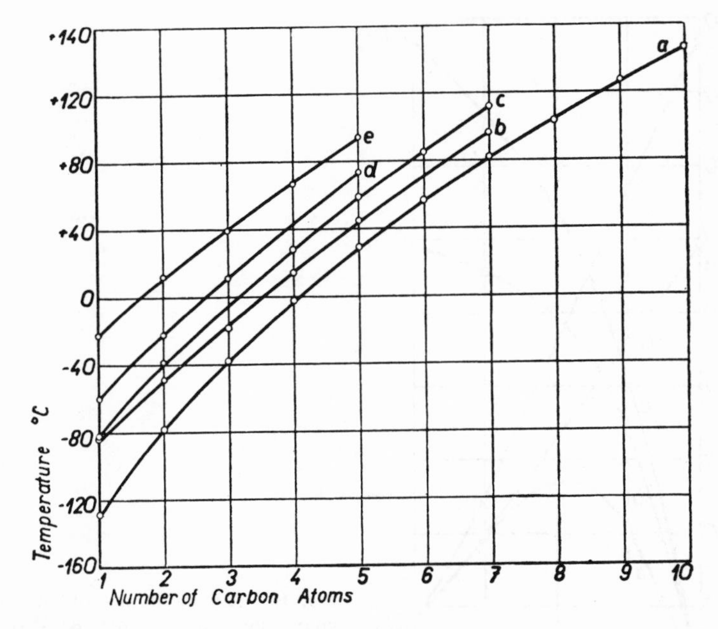


**Fig. 9.** Boiling points of (a) perfluoroparaffins, (b) perfluoroalkyl hydrides, (c) perfluoroalkyl chlorides, (d) perfluoroalkyl bromides, and (e) perfluoroalkyl iodides.

ants and monomers are available in leaflets and circulars of individual producers. Vapor pressure–temperature curves of the most common fluorinated hydrocarbons and halohydrocarbons are shown in Fig. 12.

Figure 13 shows a remarkable parallelism between the boiling points of carboxylic and perfluorocarboxylic acids. Perfluorinated acids boil systematically lower by approximately 50°C than the parent compounds.

**Table 19. Effect of Isomerism on the Melting Points and Boiling Points of Fluorohaloethanes**

| Isomer | M.P., °C | B.P., °C | Isomer | M.P., °C | B.P., °C |
|---|---|---|---|---|---|
| $CH_2FCH_2F$ | Liq. | 10.5 | $CHF_2CH_3$ | Liq. | −24.7 |
| $CHF_2CH_2F$ | Liq. | 5 | $CF_3CH_3$ | Liq. | −46.7 |
| $CCl_2FCCl_2F$ | 25 | 92.8 | $CClF_2CCl_3$ | 40.6 | 91 |
| $CClF_2CCl_2F$ | −36.4 | 47.7 | $CF_3CCl_3$ | 14.2 | 45.9 |
| $CBrF_2CHClF$ | Liq. | 52 | $CF_3CHBrCl$ | Liq. | 50 |
| $CBrF_2CBrClF$ | Liq. | 92.5 | $CF_3CBr_2Cl$ | 45.5 | 92 |
| $CBrClFCBrClF$ | 32.9 | 140 | $CBrF_2CBrCl_2$ | 45.5 | 139 |

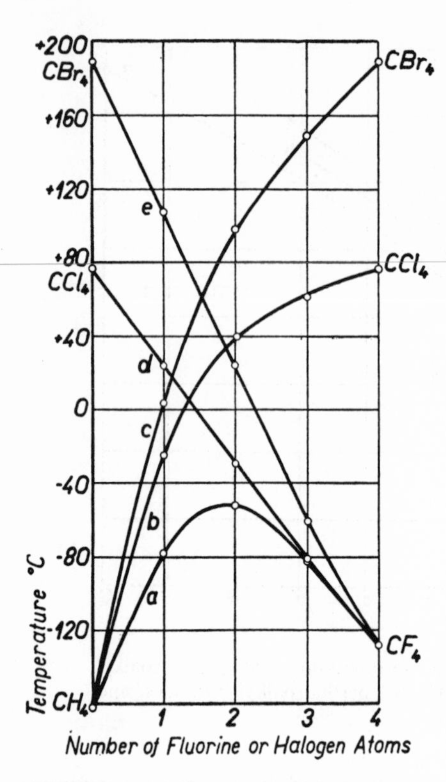


**Fig. 10.** Boiling points of fluorinated, chlorinated, and brominated derivatives of methane: (a) the series $CH_4$ — $CF_4$, (b) the series $CH_4$ — $CCl_4$, (c) the series $CH_4$ — $CBr_4$, (d) the series $CCl_4$ — $CF_4$, and (e) the series $CBr_4$ — $CF_4$.

## Density

The density of a fluorinated derivative is always higher than that of the corresponding parent compound. The most complete data are available for fluorocarbons and perfluorocarboxylic acids (Figs. 14, 15). The temperature coefficient of the density for perfluoroparaffins is $-0.0023$ to 0.0025 deg$^{-1}$ (for paraffins, approximately $-0.0008$ deg$^{-1}$).

## Refractive Index

The refractive index of a fluorinated derivative is considerably lower than that of the parent compound. Perfluoroparaffins show refractive indices lower by approximately 0.1 units than the corresponding paraffins (Fig. 14). The refractive index $n_D^{20}=1.245$ of perfluoropentane is probably the lowest value ever recorded, and special apparatus is required for such measurements. The temperature coefficient of the refractive index of perfluoroparaffins (0.0004 deg$^{-1}$) is slightly lower than that of paraffins (0.0005 deg$^{-1}$). The atomic refractivity of fluorine is 1.0–1.2 (depending on the type of compound) for Eisenlohr values, and 0.80 for Vogel's values, for the *D* line.

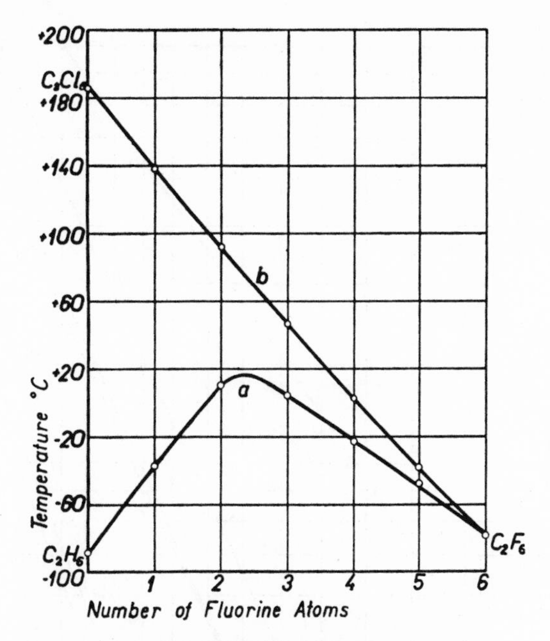


**Fig. 11.** Boiling points of fluorinated derivatives of ethane and chloroethanes: (a) the series $C_2H_6$ — $C_2F_6$, (b) the series $C_2Cl_6$ — $C_2F_6$.

## Dielectric Constant

The dielectric constant of fluorinated derivatives depends largely on the ratio of fluorine to hydrogen atoms. Perfluoroparaffins show slightly higher dielectric constants than paraffins, but polyfluoroparaffins containing a small number of hydrogen atoms possess considerably higher dielectric constants. The values given in Table 20 are illustrative of the effect of hydrogen in polyfluoro compounds.

## Surface Tension

Surface tension is one property which is remarkably affected by the number of fluorine atoms in the organic molecule. Perfluoroparaffins have the lowest recorded surface tension, ranging from 10 to 20 dyn/cm at 20°C.

**Table 20. Dielectric Constants of Fluorinated Heptanes**

| Compound | $C_8H_{18}$ [a] | $C_7F_{16}$ | $C_7HF_{15}$ | $C_7H_2F_{14}$ |
|---|---|---|---|---|
| Dielectric constant | 1.948 | 1.765 | 2.47 | 3.18 |

[a] For comparison with hydrocarbons.

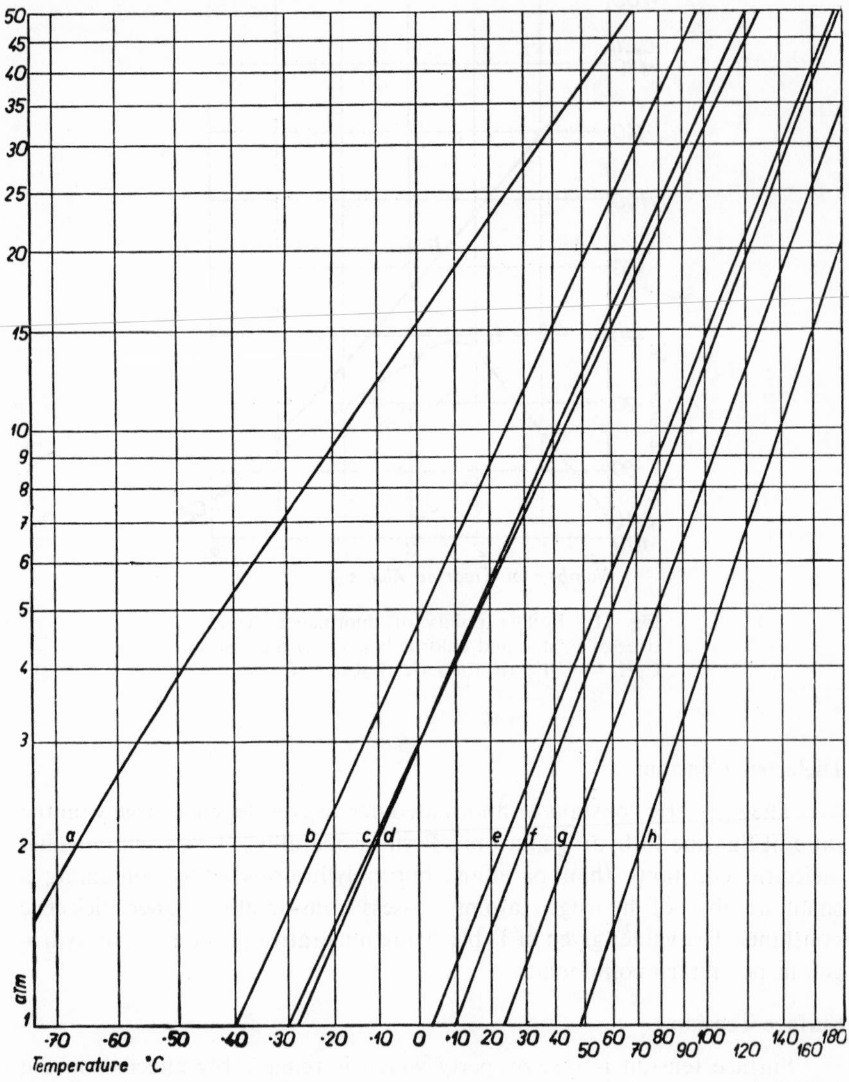


**Fig. 12.** Dependence of the vapor pressures of Freons on temperature: (a) $CClF_3$, (b) $CHClF_2$, (c) $CCl_2F_2$, (d) $C_2ClF_3$, (e) $C_2Cl_2F_4$, (f) $CHCl_2F$, (g) $CCl_3F$, (h) $C_2Cl_3F_3$.

It may take as many as 350 drops to form 1 ml of perfluoroheptane. The parachor values range from 22.2 to 26.1.

Not only do polyfluorinated compounds have very low surface tension, but some of them considerably decrease the surface tension of other com-

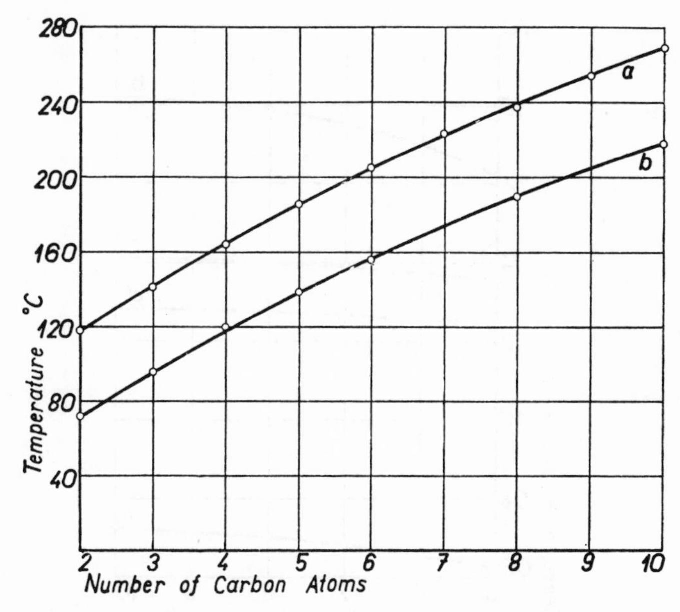


**Fig. 13.** Boiling points of (a) carboxylic acids and (b) perfluorocarboxylic aliphatic acids.

pounds. The addition of less than 1 % of perfluorocarboxylic acids to water decreases the surface tension from 72 to 20 dyn/cm.

### Viscosity

The absolute viscosity of fluorocarbons is higher and the kinematic viscosity is lower than that of the corresponding paraffins (Fig. 16). The viscosity index—the change of viscosity with temperature—is considerably higher than that of paraffins. This limits the application of poly- and perfluorinated greases and lubricants.

### Solubility

The solubility of fluorinated and perfluorinated hydrocarbons in water is negligible, and in organic solvents is usually lower than that of the corresponding parent compounds. Very frequently, two-phase systems are formed with organic solvents.

## PHYSICOCHEMICAL PROPERTIES

The strong electronegative inductive effect of fluorine in carbon chains decreases the *basicity* of fluorinated amines (Table 21) **[209]**, and increases

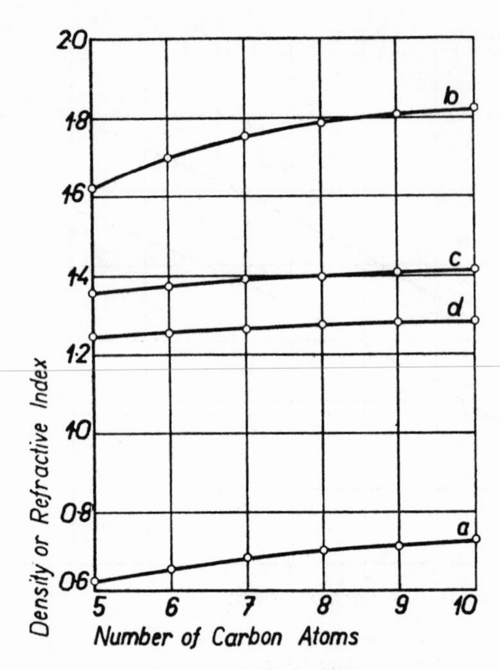


**Fig. 14.** (a,b) Densities and (c,d) refractive indices of paraffins and perfluoroparaffins, respectively.

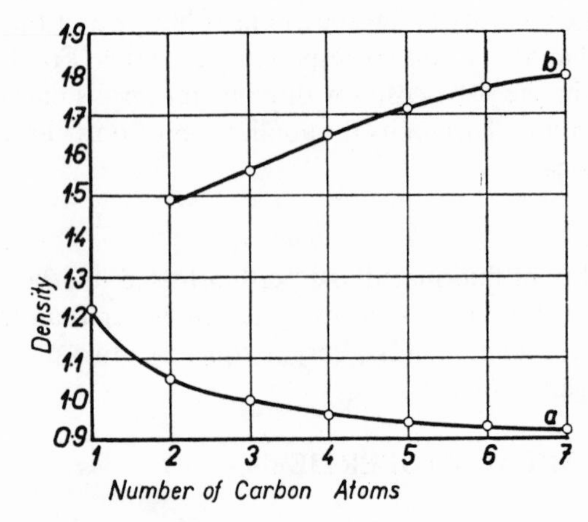


**Fig. 15.** Density of (a) carboxylic acids and (b) perfluorocarboxylic acids.

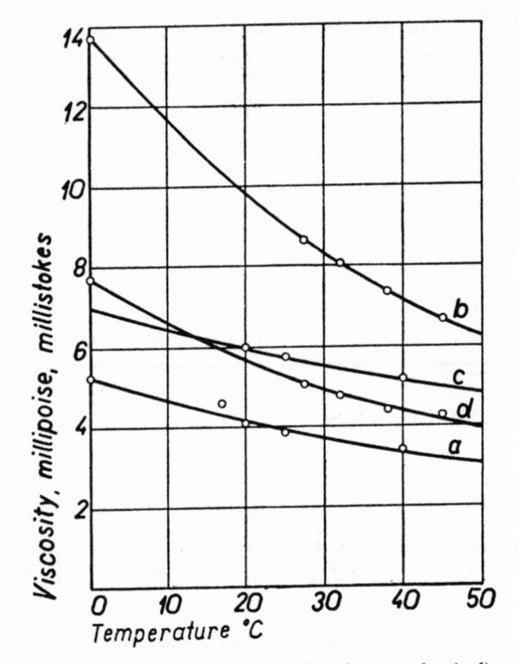


**Fig. 16.** (a,b) Absolute viscosity and (c,d) kinematic viscosity of heptane and perfluoroheptane, respectively.

considerably the *acidity* of fluorinated alcohols, phenols, and acids. Perfluorinated alcohols exhibit the acidity of phenols (Table 22) **[210–212]**, pentafluorophenol ($K_a = 4.79 \times 10^{-6}$) **[213]** the acidity of organic acids, and perfluorinated carboxylic acids the acidity of mineral acids (Table 23) **[214]**. Perfluoroalkylphosphonic acids even exceed in acidity common mineral acids and compare with the acidity of perchloric acid (Table 24) **[215]**.

Judging from the velocity constants of hydrogen–deuterium exchange, single fluorine atoms in perfluoroparafins show strong acidity. This acidity increases with the increasing number of difluoromethylene or perfluoroalkyl groups flanking the hydrogen-carrying carbon atom (Table 25) **[216]**.

**Table 21. Dissociation Constants of Fluorinated Amines [209]**

| Amine | $K_b$ | Fluorinated amine | $K_b$ |
|---|---|---|---|
| $CH_3CH_2NH_2$ | $4.5 \times 10^{-4}$ | $CF_3CH_2NH_2$ | $5 \times 10^{-9}$ |
| $CH_3CH_2CH_2NH_2$ | $4.5 \times 10^{-4}$ | $CF_3CH_2CH_2NH_2$ | $5 \times 10^{-6}$ |

**Table 22. Dissociation Constants of Fluorinated Alcohols**

| Fluorinated alcohol | Dissociation constant $K_a 25$ | p$K_a$ | Ref. |
|---|---|---|---|
| $CF_3CH_2OH$ | $4.0 \times 10^{-12}$ | 11.4 | 210 |
| | $4.0 \times 10^{-13}$ | 12.4 | 211 |
| $(CF_3)_2CHOH$ | $5.0 \times 10^{-10}$ | 9.3 | 211 |
| $(CF_3)_3COH$ | $3.0 \times 10^{-10}$ | 9.5 | 212 |
| | $6.3 \times 10^{-6}$ | 5.2 | 211 |
| $C_3F_7CH_2OH$ | $4.3 \times 10^{-12}$ | 11.4 | 210 |
| $(C_3F_7)_2CHOH$ | $2.2 \times 10^{-11}$ | 10.7 | 210 |
| $(C_3F_7)_3COH$ | $1.0 \times 10^{-10}$ | 10.0 | 212 |

## BIOLOGICAL PROPERTIES

A great many fluorinated compounds show biological activity, and the number of compounds prepared with this goal in mind is ever increasing. The most remarkable biological property of some fluorinated compounds is the effect on enzymes. Indeed, fluorophosphates and fluoroacetates are the most efficient enzymic poisons.

Fluorophosphates and kindred compounds inhibit the enzyme cholinesterase which hydrolyzes acetylcholine to choline. In this way, transfer of nerve impulses to certain muscles is interrupted.

The main representative of this type of compound is *diisopropylfluorophosphate* (DFP), which is used for treating glaucoma. Two more compounds, isopropyl and *sec*-neohexyl methanefluorophosphonates, are known from warfare as the "nerve poisons" Sarin and Soman, respectively (p. 78) **[217]**.

Another group of physiologically active compounds comprises *fluoroacetic acid*, its derivatives, and compounds which can be biologically degraded to fluoroacetic acid **[218]**. A common feature of all these com-

**Table 23. Dissociation Constants of Fluorinated Carboxylic Acids [214]**

| Fluorinated acid | Dissociation constant $K_{25}$ |
|---|---|
| $CH_3CO_2H$ | $1.8 \times 10^{-5}$ |
| $CH_2FCO_2H$ | $2.2 \times 10^{-3}$ |
| $CHF_2CO_2H$ | $5.7 \times 10^{-2}$ |
| $CF_3CO_2H$ | $5.9 \times 10^{-1}$ |
| $CF_3CH_2CO_2H$ | $0.95 \times 10^{-3}$ |
| $CF_3CH_2CH_2CO_2H$ | $7 \times 10^{-5}$ |
| $CF_3CH_2CH_2CH_2CO_2H$ | $3.2 \times 10^{-5}$ |
| $CH_3CH_2CH_2CH_2CO_2H$ | $1.6 \times 10^{-5}$ |

Table 24. Relative Acidities of Mineral Acids and Fluorinated Organic Acids [215]

| Acid | Relative acidity |
|---|---|
| $HNO_3$ | 1 |
| $CF_3CO_2H$ | 1 |
| $C_3F_7CO_2H$ | 1 |
| $CF_3As(O)(OH)_2$ | 2.5 |
| $(CF_3)_2As(O)OH$ | 3.5 |
| HCl | 9 |
| $CF_3P(O)(OH)_2$ | 9 |
| $H_2SO_4$ | 32 |
| HBr | 180 |
| $(CF_3)_2P(O)OH$ | 250 |
| $HClO_4$ | 360 |

pounds is the ultimate biological conversion to fluorocitric acid, which blocks the enzyme aconitase **[219]**. As a consequence, the Krebs tricarboxylic acid cycle is stopped at the stage of citric acid, which is not degraded, accumulates in certain organs, and causes paralysis of muscles.

The story of the discovery of these properties is one of the most exciting in fluorine chemistry. It is closely connected with the isolation of two fluorinated natural products. One is potassium fluoroacetate, which is the toxic principle of a South African plant, *Dichapetalum cymosum* (gifblaar) **[217,218]**, and of an Australian plant, *Acacia georginae* (gidyea) **[220]**. The other is fluoroöleic acid, which was isolated from another South African plant, *Dichapetalum toxicarium* (ratsbane) **[221]**. $\omega$-Fluoroöleic acid is easily degraded in the organism to fluoroacetic acid. Other compounds containing fluorine in the $\omega$-position of an aliphatic chain show the same type of toxicity as long as they can be degraded to fluoroacetic acid. Since

Table 25. Relative Acidity of Hydrogen Based on Hydrogen–Deuterium Exchange Rate [216]

| Compound | Relative acidity of hydrogen |
|---|---|
| $CHF_3$ | 1 |
| $CHF_2(CF_2)_5CF_3$ | 6 |
| $CHF(CF_3)_2$ | $2 \times 10^5$ |
| $CH(CF_3)_3$ | $1 \times 10^9$ |
| CH(–$CF_2CF_2$–)(–$CF_2$–)(–$CF_2CF_2$–)CF (bicyclic: CH and CF bridgeheads linked by $CF_2CF_2$, $CF_2$, $CF_2CF_2$ bridges) | $5 \times 10^9$ |

**Table 26. Toxicities of Alkyl ω-Fluorocarboxylates Tested in Mice by Injections in Propylene Glycol Solutions [218]**

| Even-number ω-fluorocarboxylate | $LD_{50}$, mg/kg | Odd-number ω-fluorocarboxylate | $LD_{50}$, mg/kg |
|---|---|---|---|
| $FCH_2CO_2CH_3$ | 15 | $F(CH_2)_2CO_2C_2H_5$ | >200 |
| $F(CH_2)_3CO_2CH_3$ | Toxic | $F(CH_2)_4CO_2C_2H_5$ | >160 |
| $F(CH_2)_5CO_2C_2H_5$ | 4 | $F(CH_2)_{10}CO_2C_2H_5$ | >100 |
| $F(CH_2)_7CO_2C_2H_5$ | 9 | — | — |
| $F(CH_2)_9CO_2C_2H_5$ | <10 | — | — |
| $F(CH_2)_{11}CO_2C_2H_5$ | 20 | — | — |
| $FCH_2C(CH_3)_2CH_2CO_2C_2H_5$ | Nontoxic | — | — |

the degradation in the organism follows the pattern of β-oxidation, only even-number carbon-atom acids (or the corresponding alcohols or aldehydes) show the toxicity. Odd-number carbon-atom acids, which do not degrade to fluoroacetic acid, are relatively nontoxic. The comparison of the two series of compounds is in Table 26.

Among fluorohalohydrocarbons which were for a long time considered as nontoxic and physiologically inert, quite a few in fact exhibit considerable toxicity. One of the most dangerous is *perfluoroisobutylene*, which has a toxicity comparable to that of phosgene. Since it is one of the products of pyrolysis of polytetrafluoroethylene, which is widely used in households, laboratories, and factories, it is advisable to be careful in handling polytetrafluoroethylene objects at temperatures higher than 350°C. Toxicity data on some fluorinated hydrocarbons are listed in Table 27 **[222,223]**.

Several fluorinated compounds show *anesthesiological properties* and are used in inhalation anesthesia, and quite a few in other fields of medicine. Such compounds will be discussed later (p. 77).

**Table 27. Toxicities of Gaseous and Liquid Fluorohalohydrocarbons [223]**

| Compound | Lethal concentration, ppm |
|---|---|
| $CH_2{=}CHF$ | 800,000 |
| $CBrF_3$ | 500,000 |
| $CCl_2F_2$ | 100,000 |
| $CF_2{=}CF_2$ | 40,000 |
| $CCl_3F$ | 25,000 |
| $CClF{=}CF_2$ | 4,000 |
| $CF_3CH{=}CClCF_3$ | 3 |
| $(CF_3)_2C{=}CF_2$ | 0.5 |

*Chapter 7*

# Practical Applications of Organic Fluorine Compounds

The development of fluorine chemistry is closely related to practical applications of fluorine derivatives. It was not until the early 1930's that fluorine chemistry became of interest to large-scale producers, after the discovery of fluorinated refrigerants commonly called "Freons." Later on, fluorinated plastics added to the importance of fluorine derivatives. The use of fluorocarbons in the field of atomic energy meant a further extension of practical uses of fluorine derivatives. And when the boom of industrial applications started to level off, several pharmaceuticals were added to the list of large-scale products. These are the most important industrial applications **[224, 225]**, and will be dealt with in sequence.

## REFRIGERANTS, PROPELLANTS, AND FIRE EXTINGUISHERS

Chlorofluoro derivatives of methane and ethane were found to possess remarkable thermodynamic properties for the purposes of *refrigeration* **[226]**. Since in addition to this they have remarkable thermal and chemical stability and physiological inertness, they gradually replaced conventional refrigerants in almost all fields of application. Proper combinations of different halogen atoms in the derivatives of methane and ethane lead to compounds of boiling points varying over a wide range of temperatures. Consequently, refrigerants suitable for household refrigerators, air-conditioning units, and deep freezers can be obtained from the same starting material. The most common refrigerants are produced from carbon tetrachloride, chloroform, and hexachloroethane (or a mixture of tetrachlorethane and chlorine) by a process using anhydrous hydrogen fluoride and catalysts antimony pentachloride for the liquid-phase process **[227]**, and ferric chloride **[228]**, chromium oxyfluoride **[229]**, or thorium tetrafluoride **[230]** for vapor-phase process.

$$CCl_4 \xrightarrow{HF} \underset{F11}{CCl_3F} + \underset{F12}{CCl_2F_2} + \underset{F13}{CClF_3}$$

(94) $$CHCl_3 \xrightarrow{HF} \underset{F21}{CHCl_2F} + \underset{F22}{CHClF_2} + \underset{F23}{CHF_3}$$

$$C_2Cl_6 \xrightarrow{HF} \underset{F111}{CCl_2FCCl_3} + \underset{F112}{CCl_2FCCl_2F} + \underset{F113}{CClF_2CCl_2F} + \underset{F114}{CClF_2CClF_2}$$

When the consumption of refrigerants started to level off and failed to keep pace with increased production facility, other uses of fluorinated derivatives of hydrocarbons were looked for. Mixtures of dichlorodifluoromethane and fluorotrichloromethane, as well as some other combinations, were found to be ideal *propellants* in dispersing herbicides, pesticides, and insecticides. Furthermore, some of them are used in cosmetic packaging for dispensing shaving creams, hair sprays, deodorants, lacquers, perfumes, etc. Perfluorocyclobutene found application in the food industry **[224,225]**.

Finally, large amounts of fluorinated methane and ethane derivatives are used for the production of *fire extinguishers* **[231]**. Dibromodifluoromethane and especially bromotrifluoromethane prepared by halogen fluorine exchange, or bromination of difluoro- or trifluoromethane, respectively, possess excellent fire-extinguishing properties and are used in fighting large-scale fires in the aerospace and fuel industries. In this respect, 1,2-dibrometetrafluoroethane is even better, but much more expensive **[232]**.

(95) $$CBr_4 \xrightarrow{HF} CBr_2F_2,\ CBrF_3$$

$$CH_2F_2 \xrightarrow{Br_2} CBr_2F_2 \qquad CBrF_3 \xleftarrow{Br_2} CHF_3$$

(96) $$CHClF_2 \xrightarrow{650°C} CF_2{=}CF_2 \xrightarrow{Br_2} CF_2BrCF_2Br$$

Trade marks of refrigerants, their physical properties, and the vapor pressures of some of them are listed in Tables 28, 29, and 30, respectively.

## PLASTICS AND ELASTOMERS

Plastics constitute only about 3–5% of the total production of fluorine compounds. However, their importance is much greater than this might indicate. Polytetrafluoroethylene (Teflon), discovered by Plunket **[233]** in 1940, remains unsurpassed as far as chemical and thermal resistance goes. Its disadvantage is difficult workability and processing, since this material cannot be molded in the conventional way. It requires preforming by com-

**Table 28. List of Trademarks and Products of Refrigerants [224]**

| Trademark | Manufacturer | Place | Country |
|---|---|---|---|
| Algofren | Montecatini | Milan | Italy |
| Algeon | Fluoder | Lanus | Argentina |
| Arcton | Imperial Chemical Industries Ltd. | Manchester | England |
| Col-flon | Nitto Chemicals Co., Ltd. | Tokyo | Japan |
| Daiflon | Osaka Metal | Tokyo | Japan |
| Edifren | Societa Edison | Milan | Italy |
| Eskimon | — | — | USSR |
| Flugene | Pechiney | Paris | France |
| Fluogen | Chemische Fabrik von Heyden | Munich | Germany |
| Fluorion | Productos Quimicos Industrial Comas Ing. | Madrid | Spain |
| Forane | Société d'électrochimie, d'électrometallurgie et des aciéries d'Ugine | Lyon | France |
| Fration | La Fluorhidrica | Buenos Aires | Argentina |
| Freon | E.I. du Pont de Nemours & Co. | Wilmington, Delaware | USA |
| | Ducilo | Buenos Aires | Argentina |
| | Du Pont do Brazil | Sao Paolo | Brazil |
| | Du Pont of Canada | Montreal | Canada |
| | Halocarburos | Mexico City | Mexico |
| Fresane | Uniechemie N.V. | Apeldoorn | Netherlands |
| Frigedohn | VEB Alcid Fluorwerke Dohna | Radebeul bei Dresden | Germany |
| Frigen | Farbwerke Hoechst A. G. | Frankfurt am Main | Germany |
| | Fongra Produtos Quimicos | Sao Paolo | Brazil |
| Genetron | Allied Chemical Corporation | New York, New York | USA |
| Isceon | Imperial Smelting Corporation Ltd. | Avonmouth | England |
| Isotron | Pennsalt Chemicals Corporation | Philadelphia, Pa. | USA |
| Ledon | Spolek pro chemickou a hutní výrobu | Ústí nad Labem | Czechoslovakia |
| Ucon | Union Carbide Corporation | New York, New York | USA |

pression and subsequent sintering. For this reason, another fluorinated polymer, poly(chlorotrifluoroethylene) (Kel *F*) temporarily gained ground. It is slightly inferior to Teflon both in thermal and chemical resistance. Recently, the processing of Teflon has been improved, and in addition, several fluorinated copolymers have been discovered. They possess remark-

**Table 29. Physical and Thermodynamic Constants of Freons**

| Name | Formula | M.P., °C | B.P., °C | Heat of evaporation at B.P., kcal/mole | Specific heat, kcal/mole | Critical temp., °C | Critical pressure, atm | Critical density, kg/liter |
|---|---|---|---|---|---|---|---|---|
| Freon 11 | $CCl_3F$ | −111.1 | 23.77 | 43.51 | 0.208 | 198.0 | 43.2 | 0.554 |
| Freon 12 | $CCl_2F_2$ | −155 | −29.8 | 39.9 | 0.204 | 111.5 | 40.95 | 0.555 |
| Freon 13 | $CClF_3$ | −180 | −81.5 | 35.65 | 0.203 | 28.8 | 39.4 | 0.581 |
| Freon 21 | $CHCl_2F$ | −135 | 8.92 | 57.85 | 0.246 | 178.5 | 51.0 | 0.522 |
| Freon 22 | $CHClF_2$ | −160 | −40.8 | 55.9 | 0.265 | 96.0 | 48.7 | 0.525 |
| Freon 23 | $CHF_3$ | −163 | −82.2 | 56 | 0.28 | 32.3 | 50.5 | — |
| Freon 113 | $C_2Cl_3F_3$ | − 35 | 47.57 | 35.04 | 0.226 | 214.1 | 33.7 | 0.576 |
| Freon 114 | $C_2Cl_2F_4$ | − 94 | 3.55 | 32.8 | 0.232 | 145.7 | 32.1 | 0.582 |
| Freon 115 | $C_2ClF_5$ | −106 | −38 | — | — | — | — | — |

able resistance and can be worked up in a conventional way. Some of them have elastic properties and resemble rubber.

The manufacture of plastics consists of three distinct stages: production of monomers, polymerization, and processing of the polymer.

## Monomers

Fluorinated monomers must be prepared in very high purity, usually at least 99.99%. In order to prevent spontaneous undesirable polymerization, stabilizers such as phenyl-$\beta$-naphthylamine or terpene B are added to the pure compounds. Fresh distillation and adsorption of stabilizer on silica gel usually precede polymerization. A survey of the most common monomers is shown in the following equations; their physical constants are given in Table 31 **[234–242]**.

**Table 30. Vapor Pressure of Freons (Decimal log, abs. atm., abs. temperature)**

| | |
|---|---|
| Freon 12: | $\log p = 1.36315 - \left(\frac{1816.5}{T}\right) - 10.859 \log T + 0.007175T$ |
| Freon 11: | $\log p = 34.8838 - \left(\frac{2303.95}{T}\right) - 11.7406 \log T + 0.0064249T$ |
| Freon 21: | $\log p = 38.2974 - \left(\frac{2367.41}{T}\right) - 13.0295 \log T + 0.0071731T$ |
| Freon 22: | $\log p = 25.1144 - \left(\frac{1638.32}{T}\right) - 8.1418 \log T + 0.0051838T$ |
| Freon 113: | $\log p = 29.5335 - \left(\frac{2406.0}{T}\right) - 9.2635 \log T + 0.0036970T$ |

**Table 31. Properties of Fluorinated Monomers Used for Plastics**

| Monomer | M.P., °C | B.P., °C | $d_4^t$ | $n_D^t$ | Critical temp., °C |
|---|---|---|---|---|---|
| Vinyl fluoride | −160.5 | −72.2 | 0.675/26°C | — | 54.7 |
| Vinylidene fluoride | −144 | −85.7 | 0.659/21°C | — | 29.7 |
| Chlorotrifluoro- | −157.5 | −26.2 | 1.51/−40°C | — | — |
| ethylene | −157.9 | −27.9 | 1.38/0°C | — | — |
| Tetrafluoroethylene | −142.5 | −76.3 | 1.533/−80°C | — | 33.3 |
| | | | 1.1507 | — | — |
| Perfluoropropylene | −158.1 | −28.2 | 1.583/−40°C | — | — |
| Perfluorobutadiene | −132.1 | 6.0 | 1.553/−20°C | 1.378/−20°C | — |
| 1,1-Dihydroperfluoro-butyl acrylate | — | 51.3/50 mm | 1.409/20°C | 1.3317/20°C | — |
| 1,1-Dihydroperfluoro-hexyl acrylate | — | 63.5/20 mm | 1.54/20°C | 1.3296/20°C | — |

(97) $$CHClF_2 \xrightarrow{650°C} (CF_2) \longrightarrow CF_2{=}CF_2$$ [234]

(98) $$CCl_2FCClF_2 \xrightarrow{Zn} CClF{=}CF_2$$ [235]

(99) $$CF_2{=}CF_2 \xrightarrow{650°C} CF_2{=}CFCF_3$$ [236]

(100) $$CF_2{=}CFCF_3 \xrightarrow{H_2/Pd} CHF_2CHFCF_3 \xrightarrow{alkali} CHF{=}CFCF_3$$ [237]

(101) $$CH{\equiv}CH \xrightarrow{HF} CH_2{=}CHF$$ [238]

(102) $$CHCl{=}CCl_2 \xrightarrow{HF} CH_2ClCClF_2 \xrightarrow{Zn} CH_2{=}CF_2$$ [239]

(103) $$C_3F_7CH_2OH \xrightarrow[CH_2{=}CHCOCl]{CH_2{=}CHCO_2H,\ (CF_3CO)_2O} C_3F_7CH_2OCOCH{=}CH_2$$ [240]

(104) $$CF_3CH{=}CH_2 \xrightarrow{CH_3SiHCl_2} CF_3CH_2CH_2Si(Cl)_2CH_3 \xrightarrow[NaOH]{H_2O} \left[ CF_3CH_2CH_2\overset{|}{\underset{|}{Si}}(O{-})CH_3 \right]_3$$ [241]

(105) $$(CF_3CO)_2O \xrightarrow{N_2O_3} CF_3COONO \xrightarrow{heat} CF_3NO$$ [242]

**Table 32. Copolymers of Various Monomers**

| Components | $CF_2:CF_2$ | $CClF:CF_2$ | $CH_2:CF_2$ | $CF_2:CFCF_3$ | $CHF:CFCF_3$ | $CF_3NO$ |
|---|---|---|---|---|---|---|
| $CF_2:CF_2$ | Teflon | — | — | Teflon 100 | — | Nitroso-rubber |
| $CClF:CF_2$ | — | Kel *F* | Kel *F* elastomer 3700 | — | — | — |
| $CH_2:CF_2$ | — | Kel *F* elastomer 5500 | Kynar | Viton | Tecnoflon | — |
| $CF_2:CFCF_3$ | Teflon 100 | — | Viton | — | — | — |
| $CHF:CFCF_3$ | — | — | Tecnoflon | — | — | — |
| $CF_3NO$ | Nitroso-rubber | — | — | — | — | — |

## Polymerization

The polymerization of fluorinated monomers, particularly chlorotrifluoroethylene and tetrafluoroethylene, in chloroform solutions using high concentrations (up to 5%) of dibenzoyl peroxide gives low-molecular-weight oils and greases which are used as inert liquids and lubricants **[243]**.

High-molecular-weight polymers are prepared by emulsion or suspension polymerization using ternary peroxide–sulfite–metal systems such as potassium persulfate–sodium bisulfite–silver nitrate **[244]**, or *tert*-butyl perbenzoate–sodium bisulfite–ferrocitrate **[245]**. Similarly, copolymerization is achieved when two or more monomers are polymerized at the same time. Examples of various combinations are shown in Table 32.

## Processing of Polymers

Most of the fluorinated polymers can be processed in the conventional ways: coagulation of the emulsion (latex), extrusion, injection molding, or compression molding. Some of polymers are subjected to aftertreatment of the finished articles by cross-linking (vulcanization), which converts linear chains to spatial nets. This process can be achieved using metal oxides such as zinc oxide and organic bases such as hexamethylene tetramine or methylene-bis(*p*-phenylene)-isocyanate. Polytetrafluoroethylene is first compressed and then sintered at a temperature of about 327°C.

Physical properties and trademarks of fiuorinated plastics and elastomers are listed in Tables 33 and 34, respectively.

## Applications of Plastics and Elastomers

*Polyvinyl fluoride* forms resistant films and coatings especially useful in preventing the weathering of buildings.

**Table 33. Comparison of the Properties of Fluorinated Polymers with Polyethylene**

| Property | Polyethylene | Polychloro-trifluoroethylene | Polytetra-fluoroethylene | Fluoroethylene–propylene copolymer | Vinylidene fluoride–perfluoropropylene copolymer |
|---|---|---|---|---|---|
| Density | 0.92 | 2.11–2.13 | 2.14–2.3 | 2.12–2.17 | 1.9 |
| Refractive index | 1.51 | 1.43 | 1.35 | 1.338 | — |
| Dielectric constant ($10^3$ cycles) | 2.3 | 2.8 | 2.1 | 2.1 | — |
| Loss factor ($10^3$ cps) | <0.0005 | 0.024 | 0.0002 | 0.0003 | — |
| Specific resistivity, ohms/cm | — | $5 \times 10^{17}$ | $>10^{18}$ | $>10^{18}$ | $5 \times 10^{12-14}$ |
| Transition temperature, 0°C | — | 212–214 | 327 | — | — |
| Tensile strength, kg/cm² | 105–125 | 400 | 175–280 | 190–220 | 130–170 |
| Tensile strength after orientation kg/cm² | — | 2100 | 1050 | — | — |
| Elongation, % | — | — | — | 250–330 | 100–700 |
| Processability, temperature, °C | — | 230–290 | 370–390 | — | 150–200 |
| Compression molding, temperature, °C | 135–150 | 230–260 | — | 310–400 | 150–200 |
| Pressure, atm | 14 | 35–1050 | 140 | — | — |

Table 34. Trademarks of Fluorinated Plastics and Elastomers

| Monomer | Trademark | Producer | Country |
|---|---|---|---|
| $CF_2{=}CF_2$ | Algaflon | Montecatini | Italy |
| | Fluon | Imperial Chemical Industries | Great Britain |
| | Ftoroplast 4F | — | USSR |
| | Teflon, TFE | E.I. du Pont de Nemours and Co. | USA |
| $CF_2{=}CClF$ | Fluorothene | Union Carbide Carbon Corp. | USA |
| | Genetron Plastic HL | General Chemicals | USA |
| | | Allied Chemical | USA |
| | Hostaflon | Farbwerke Hoechst | Germany |
| | Kel *F* | Minnesota Mining and Manufacturing Co. | USA |
| | Polyfluoron | Acme Resin Corp. | USA |
| | Teflex | Spolek pro chemickou a hutní vyrobu | Czecho-slovakia |
| $CH_2{=}CF_2$ | Kynar | Pennsalt Co. | USA |
| $CH_2{=}CHF$ | Dalvor | Diamond Alkali | USA |
| | Tedlar | E.I. du Pont de Nemours and Co. | USA |
| $CH_2{=}CF_2{+}CF_2{=}CClF$ | Aclar, Halon | General Chemicals | USA |
| | Kel-*F* elastomer | Minnesota Mining and Manufacturing Co. | USA |
| $CH_2{=}CF_2{+}CF_2{=}CFCF_3$ | Fluorel | Minnesota Mining and Manufacturing Co. | USA |
| | Viton | E.I. du Pont de Nemours and Co. | USA |
| $CF_2{=}CF_2{+}CF_2{=}CFCF_3$ | Teflon 100, FEP resin | E.I. du Pont de Nemours and Co. | USA |
| $CH_2{=}CF_2{+}CHF{=}CFCF_3$ | Tecnoflon | Montecatini | Italy |
| $CH_2{=}CHCOOCH_2C_3F_7$ | Fluororubber 1F4 | Minnesota Mining and Manufacturing Co. | USA |
| $-[CF_3CH_2CH_2\overset{\vert}{\underset{\vert}{Si}}CH_3-O]-$ (vertical bond above Si; Si–O–, with O below Si) | Silastic LS 53 | Dow Corning Co. | USA |

*Poly(chlorotrifluoroethylene)* is used for tubing, chemical vessels, moldings, gaskets, etc. It resists most chemicals and stands temperatures up to 180°C.

*Polytetrafluoroethylene* resists temperatures up to 250°C and all chemicals with the exception of fluorine at higher temperatures and molten sodium. Consequently, it has the broadest field of application: tubing, containers, valves, stopcocks, tapes, coatings, self-lubricating bearings, filaments, pads, and kitchen utensils. Some of these applications are due to some remarkable properties of polytetrafluoroethylene: nonwettability by aqueous as well as organic liquids and self-lubrication due to greasy surface.

*Copolymers of tetrafluoroethylene and perfluoropropylene* (PFEP) combine the strong thermal and chemical resistance of polytetrafluoroethylene with the workability of less-fluorinated plastics. They are especially useful for injection-molding technique.

Some of the copolymers of *vinylidene fluoride with chlorotrifluoroethylene or perfluoropropylene* show elastic properties, especially at low temperatures. For this purpose, *tetrafluoroethylene–trifluoronitrosomethane copolymer* also is suitable material. *Fluorinated polysiloxanes*, on the other hand, are elastomers suitable for use at higher temperatures.

Finally, *acrylates of pseudoperfluoro alcohols* are used for the production of heat-resistant textiles.

More information about the fluorinated plastics can be found in the review literature **[225]**.

## FLUORINATED COMPOUNDS AS PHARMACEUTICALS

The biological properties of many fluorinated compounds account for their use in medicine **[246]**.

The most useful applications of fluorinated pharmaceuticals are in the field of *inhalation anesthetics*. 1-Bromo-1-chloro-2,2,2-trifluoroethane (Halothane, Fluothane) has now entirely replaced ether in many hospitals. The main advantages of Halothane are its nonexplosiveness, nonflammability, and high efficiency with no postnarcotic effects. Different ways of preparation of this compound are illustrated below:

(106) $$CCl_2{=}CHCl \xrightarrow{HF} CF_3CH_2Cl \xrightarrow[500°C]{Br_2} CF_3CHClBr \qquad \text{ICI}$$

(107) $$CClF{=}CF_2 \xrightarrow{HBr} CHClFCF_2Br \xrightarrow{AlCl_3} CF_3CHClBr \qquad \text{Hoechst}$$

(108) $$CClF{=}CF_2 \xrightarrow{Br_2} CClFBrCF_2Br \xrightarrow{AlCl_3} CF_3CClBr_2 \xrightarrow[NaOH]{Na_2SO_3} CF_3CHClBr \qquad \text{Dohna}$$

(109) $$\overset{*}{C}HClBrCO_2H \xrightarrow[20°C]{SF_4} \overset{*}{C}HClBrCF_3$$

Another inhalation anesthetic, 2,2-dichloro-1,1-difluoroethyl methyl ether (Methoxyflurane, Penthrane) is not as efficient as Halothane, but does not require such severe precautions.

(110) $CCl_3CClF_2 \xrightarrow{Zn} CCl_2{=}CF_2 \xrightarrow[KOH]{CH_3OH} CHCl_2CF_2OCH_3$

Diisopropylfluorophosphate is used in the treatment of glaucoma (Neoglaucit). Two similar compounds, Sarin and Soman, can hardly be classified within the category of pharmaceuticals, since they were designed as "nerve poisons" in chemical warfare.

(111) $[(CH_3)_2CHO]_2P(O)F$ Neoglaucit; $CH_3P(O)(F)OCH(CH_3)_2$ Sarin; $CH_3P(O)(F)OCH(CH_3)C(CH_3)_3$ Soman

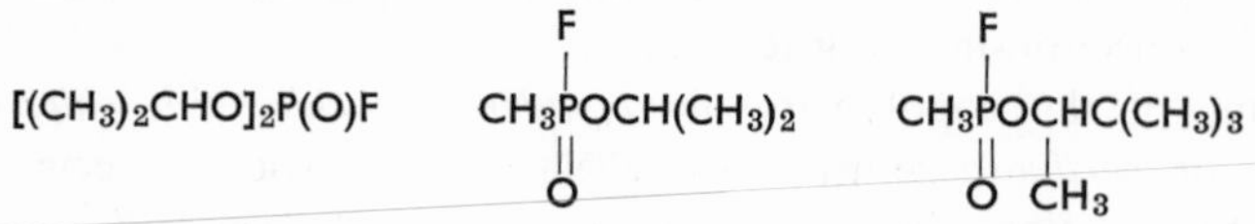

Of the many *fluorinated steroids*, those corticoids containing fluorine in positions 6, 9, or both found applications as antiinflammatory drugs and preparations with glucocorticoid activity. Triamcinolone, Dexamethasone, Paramethasone, and Synalar are the best known representatives of this group.

(112) [structures: COCH₂OH, HO, OH, F substituents]

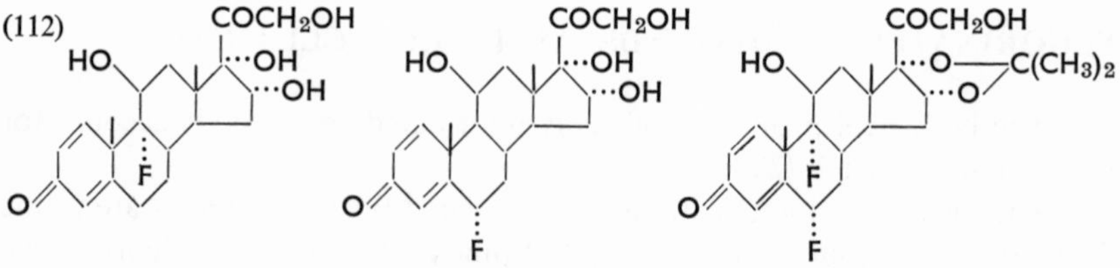

Triamcinolone Parametasone Synalar

Fluorinated phenothiazine derivatives such as Vesprin—2-trifluoromethyl-10-(3-dimethylaminopropyl)phenothiazine—are used as *tranquilizers*, and 5-fluorouracil and 5-fluoro-4-hydroxyprimidine (Fluoxidin) for *treatment* of certain forms *of cancer*.

(113) [structures: $CH_2CH_2CH_2N(CH_3)_2$, N, S, $CF_3$; F, OH, N, N, OH; F, OH, N, N]

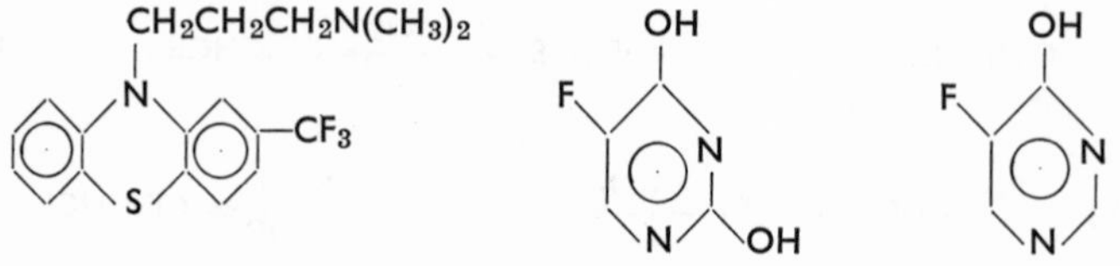

Vesprin Fluorouracil Fluoxidin

## OTHER USES OF FLUORINATED COMPOUNDS

Apart from the main applications of fluorinated derivatives as refrigerants, propellants, plastics, and pharmaceuticals already discussed, there are many other fields in which fluorinated compounds have found use **[225]**.

Higher fluorocarbons are used as *greases* and *heat transfer media* capable of resisting nocuous effects of prolonged heating. Good dielectric properties make for the use of fluorinated hydrocarbons in *electronics*.

Amides of fluorinated carboxylic and alkanesulfonic acids in which nitrogen was converted to a quaternary ion are excellent *surfactants*, decreasing surface tension in neutral, strongly acidic, and alkaline media. Some of these compounds spread over the surface of gasoline in storage tanks decrease its evaporation and inflammability.

Of similar nature are compounds that are used to form *water- and stain-repellent* films on the surface of textile fibers (Scotchgard FC 154, Zepel, Quarpel, etc.).

*Chapter 8*

# Reactions of Organic Fluorine Compounds

The factual material to be discussed in this chapter on the reactions of organic fluorine compounds is so abundant that it is out of the question to mention all reactions. The selection will therefore be limited to the most important types, and special attention will be drawn to reactions which are used for synthetic purposes and to reactions in which fluorine is eliminated from fluorinated compounds. After factoring out some common features, individual reactions will be dealt with in sequence. In order to facilitate orientation and prevent excessive overlapping, the system used here is a combination of the classical system of synthetic methods used previously in Groggin's Unit Processes with a somewhat more modern concept of mechanistic classification of reactions. There are, necessarily, flaws in this kind of arrangement, and the subject index should be used to locate the desired reaction when its position in the present system is ambiguous.

After considering reduction and oxidation, electrophilic reactions (both additions and substitutions) and nucleophilic substitutions will be discussed. They will be followed by nucleophilic and free-radical addition reactions, eliminations, rearrangements, and pyroreactions. Finally, some space will be given to reactions in which fluorinated compounds act as chemical reagents.

Since the most important reaction conditions are shown in chemical equations, which are thus self-explanatory, the amount of description in the text will be cut down to a minimum.

## FACTORS GOVERNING THE REACTIVITY OF ORGANIC FLUORINE COMPOUNDS

The reactivity of organic fluorine compounds is strongly influenced by four main factors: The inductive, hyperconjugative, mesomeric, and steric effects.

### Inductive Effect

Fluorine has a very strong inductive, electron-attracting effect. Hydrogen atoms on a carbon atom adjacent to a difluoromethylene or a

**Table 35. Velocity Constants of Displacement of Iodine by Thiophenoxide Ion at 20°C [247]**

| Alkyl iodide | $CH_3CH_2I$ | $CH_2FCH_2I$ | $CHF_2CH_2I$ | $CF_3CH_2I$ |
|---|---|---|---|---|
| $k_2 \times 10^5$ (liters/mole·sec) | 2,600 | 166 | 7.34 | 0.149 |
| Relative reactivity | 17,450 | 1,113 | 49.2 | 1 |

trifluoromethyl group are strongly acidic, as shown by easy replacement by deuterium (Table 25). The same effect causes a decrease in the basicity of fluorinated amines (Table 21) and an increase in the acidity of fluorinated alcohols (Table 22), fluorinated phenols, and fluorinated acids (Tables 23, 24) as compared with the nonfluorinated parent compounds (pp. 65–67).

The accumulation of fluorine at a carbon adjacent to a carbon-bonding iodine in alkyl iodides accounts for a strong decrease in nucleophilicity of the fluoroalkyl group **[247]** in alkylation reactions (Table 35).

The inductive effect of difluoromethylene group, trifluoromethyl groups and generally perfluoroalkyl groups is responsible for the easy formation and stability of addition compounds of fluorinated ketones with hydrogen fluoride **[248]**, water **[249]**, and ammonia **[250]**. The last reaction is the probable cause of the very low yields of the reaction of trifluoromethyl ketones with acetylene in liquid ammonia.

(114) $$\begin{array}{l} CF_2{-}CF_2 \\ |\quad\quad | \\ CF_2{-}CO \end{array} \xrightarrow{HF} \begin{array}{l} CF_2{-}CF_2 \\ |\quad\quad | \\ CF_2{-}C{-}OH \\ \quad\quad\ \ | \\ \quad\quad\ \ F \end{array}$$ **[248]**

(115) $$(C_3F_7)_2CO \xrightarrow{H_2O} (C_3F_7)_2C(OH)_2$$ **[249]**

(116) $$CF_3COCH_3 \xrightarrow{NH_3} CF_3C(NH_2)(OH)CH_3$$ **[250]**

The inductive effect of trifluoromethyl groups favors the formation of epoxides from fluorinated ketones and diazomethane **[251]**:

(117) $$(CF_3)_2CO \xrightarrow{CH_2N_2} (CF_3)_2\overset{O}{\overbrace{C{-}CH_2}}$$

The inductive effect, which may be combined with steric effects, is responsible for the large proportion of reduction as compared with addition in

**Table 36. Effect of Fluorine Substitution on the Addition/Reduction Ratio in the Reaction of Aldehydes with Ethylmagnesium Bromide [252]**

| R[a] | Effective diameter, Å | Addition, % | Reduction, % |
|---|---|---|---|
| $CH_3$ | 4.0 | 100 | 0 |
| $CF_3$ | 5.1 | 60 | 20 |

[a] Aldehyde RCHO.

the reaction of Grignard reagents to fluorinated aldehydes, ketones, esters, and nitriles **[252]** (see also p. 120) (Table 36).

The inductive effect of fluorine stabilizes hydroxyl-containing intermediates in the base-catalyzed additions across carbonyl groups **[250,253]**. Consequently, dehydration to $\alpha,\beta$-unsaturated compounds does not take place readily.

(118) $C_2H_5OCOCHFCOCO_2C_2H_5 + C_6H_5NHNH_2 \longrightarrow C_2H_5OCOCHFC(OH)(NHNHC_6H_5)CO_2C_2H_5$ **[250]**

(119) $CF_3COCH_3 + CH_2(CO_2H)_2 \longrightarrow (CF_3)(CH_3)C(OH)CH(CO_2H)_2 \longrightarrow (CF_3)(CH_3)C(OH)CH_2CO_2H$ **[253]**

By the same token, polyfluorinated alcohols are very resistant to dehydration **[254]**:

(120) $CF_3CH(OH)CH_3 \xrightarrow[180\text{–}190°C]{H_2SO_4}$ N.R.; $CF_3CH(OH)CH_3 \xrightarrow[150\text{–}200°C]{P_2O_5}$ N.R.

The inductive effect of fluorine plays an important role in substitution reactions. The Friedel-Crafts reaction of both 3-chloro- and 2-chloro-1,1,1-trifluoropropane with benzene gives the same product, 3,3,3-trifluoro-*n*-propylbenzene. This result contrasts with analogous reactions of propyl and isopropyl chloride, which both react with benzene to give isopropylbenzene. The difference is best accounted for by the inductive effect of trifluoromethyl groups, which stabilize the intermediate carbonium ions at the more distant carbon atoms **[255]**.

(121) [255]

$$CF_3CH_2CH_2Cl \xrightarrow{AlCl_3} CF_3CH_2\overset{\oplus}{C}H_2 \xrightarrow{C_6H_6} CF_3CH_2CH_2C_6H_5$$

$$CF_3CHClCH_3 \xrightarrow{AlCl_3} CF_3\overset{\oplus}{C}HCH_3 \longrightarrow CF_3CH_2\overset{\oplus}{C}H_2 \xrightarrow{C_6H_6} CF_3CH_2CH_2C_6H_5$$

The inductive effect of fluorine shows further in the drainage of electrons from the double bonds of poly- and perfluoro-olefins so that electrophilic additions are impossible or very difficult. On the other hand, these olefins react readily with nucleophiles such as marcaptans, thiophenols, phenols, alcohols, amines, and even fluoride ions (generated from potassium or cesium fluorides) **[256]**.

(122) [256]

$$\begin{matrix} CF_2{=}CF_2 \\ CF_2{=}CFCl \\ CF_2{=}CFCF_3 \end{matrix} \xrightarrow[H^{\oplus}\ (HCONH_2)]{F^{\ominus}\ (KF)} \begin{matrix} CF_3CHF_2 \\ CF_3CHFCl \\ CF_3CHFCF_3 \end{matrix}$$

Similarly, fluorine atoms in poly- and perfluoroaromatics drain electrons from the nucleus and make it susceptible to nucleophilic attacks by hydroxyl, alkoxyl, aryloxyl, sulfhydryl, and amino groups **[257,258]**.

(123) [257]

$$C_6F_6 \longrightarrow \begin{cases} C_6F_5OH,\ C_6F_5OR \\ C_6F_5SH,\ C_6F_5SR \\ C_6F_5NH_2\ C_6F_5NR_2,\ C_6F_5NHNH_2 \end{cases}$$

(124) Relative reactivities toward $\overset{\ominus}{O}CH_3$, [258]

$C_6H_5F$ 1

$C_6F_6$ $1.39 \times 10^8$

Sufficiently electron-poor systems such as perfluoropyridine or perfluoroacyl fluorides may even be attacked by carbanions produced, for example, by the reaction of perfluoro-olefins with potassium or cesium fluoride **[259,260]**.

(125) [259]

$$CF_2{=}CFCF_3 \xrightarrow[(KF)]{F^{\ominus}} CF_3\overset{\ominus}{C}FCF_3 \xrightarrow[-F^{\ominus}]{C_5F_5N} CF_3CFCF_3 \text{ (bonded to C-4 of 2,3,5,6-tetrafluoropyridine)}$$

(126) [260]

$$CF_2{=}CFCF_3 \xrightarrow[(CsF)]{F^{\ominus}} CF_3\overset{\ominus}{C}FCF_3 \xrightarrow[-F^{\ominus}]{CF_3COF} CF_3CF(COCF_3)CF_3$$

The inductive effect of perfluoroalkyl groups is further responsible for an easy nucleophilic attack by hydroxyl of perfluoroalkyl ketones in haloform-type reactions **[261]**.

$$(127)\qquad H\overset{\ominus}{O} \dashrightarrow \overset{R}{\overset{|}{C}}OC_3F_7 \longrightarrow HO\overset{R}{\overset{|}{C}}O + \overset{\ominus}{C}_3F_7 \xrightarrow{H^{\oplus}} C_3HF_7 \qquad [261]$$

## Hyperconjugation

Combined with the inductive effect, hyperconjugation can account for the anti-Markovnikov additions across the double bond in 3,3,3-trifluoropropene. Because of the concentration of negative charge at the central carbon atom, halide anions join the terminal carbon atom **[262]**. On the other hand, in the reaction of 3,3,3-trifluoropropene with bromine in acetic acid, where bromine cation is the attacking species, the terminal carbon atom combines with the acetate anion **[263]**.

$$(128)\qquad CF_3CH{=}CH_2 \longrightarrow \overset{\ominus}{F}\ CF_2{=}CH{-}\overset{\oplus}{C}H_2 \begin{cases} \xrightarrow[AlCl_3]{HX} CF_3CH_2CH_2X,\ X{=}Cl,\ Br & [262] \\ \xrightarrow[CH_3CO_2H]{Br_2} CF_3CHBrCH_2OCOCH_3 & [263] \end{cases}$$

Two trifluoromethyl groups at one carbon atom of a double bond overbalance even the strong mesomeric effect of a carboxylic group attached to the other carbon atom of the same double bond, as in the addition of ammonia to diethylperfluoroisopropylidenemalonate (one trifluoromethyl group is not strong enough to achieve the same direction of addition) **[264]**.

$$(129)\qquad (CF_3)_2C{=}C(CO_2C_2H_5)_2 \xrightarrow{NH_3} (CF_3)_2CH\underset{NH_2}{\underset{|}{C}}(CO_2C_2H_5)_2 \qquad [264]$$

The orientation to the *meta*-position in electrophilic substitutions in the benzene ring carrying a trifluoromethyl group may also be due to hyperconjugation **[265]**.

(130) [265]

CF₃-benzene ⟷ $\overset{\ominus}{F}\ CF_2$ (ortho $\oplus$) ⟷ $\overset{\ominus}{F}\ CF_2$ (para $\oplus$) $\xrightarrow{Br_2}$ m-Br-C₆H₄-CF₃

## Mesomeric Effect

The electron-releasing force of fluorine is stronger than that of the other halogen atoms in fluorohalo-olefins. Consequently, nucleophiles always join preferentially the carbon atom carrying more fluorine atoms **[266,267]**.

(131) $F(Cl)C{=}CF_2 \longrightarrow F(Cl)\overset{\ominus}{C}{-}C(F){=}\overset{\oplus}{F}$ → HBr: $CHClFCF_2Br$ **[266]**; ROH: $CHClFCF_2OR$ **[267]**

In the aromatic series, the mesomeric effect of fluorine accounts for *ortho–para* orientation in electrophilic substitutions **[268]**. This effect exceeds in power the combined inductive and hyperconjugative effects of the methyl group **[269]**. It is probably also responsible for the high reactivity of fluorodurene in bromination **[270]**.

(132) Fluorobenzene ↔ resonance forms $\xrightarrow[H_2SO_4]{HNO_3}$ p-nitrofluorobenzene (90%) + o-nitrofluorobenzene (8%) + m-nitrofluorobenzene **[268]**

(133) 2-Fluorotoluene $\xrightarrow[H_2SO_4]{HNO_3}$ 84% + 10% + 6% nitro isomers **[269]**

(134) Bromination ($Br_2$) of X-substituted durene **[270]**

| X = | H | F | Cl | Br | I |
|---|---|---|---|---|---|
| Rel. reactivity: | 1000 | 2310 | 72.6 | 30.9 | 40 |

## Steric Effects

The effective diameter of fluorine (Table 36) does not always give a clue to the reactions of fluorinated compounds. In monofluoro compounds, fluorine does not seem to occupy much more space than a hydrogen atom. On the other hand, difluoromethylene and trifluoromethyl groups show much larger steric requirements than methylene or methyl groups, respectively.

Sometimes it is difficult to distinguish the steric effect of fluorine from its inductive effect. It is questionable which of them governs the orientation of the free-radical addition of alkyl and halofluoroalkyl groups to olefins and halofluoro-olefins. The results of these reactions can be interpreted by assuming either effect as operating **[271,272]**.

(135)

$$H < F < Cl < CH_3 < CF_3 \qquad [271]$$

$$CH_2 < CHF < CHCl < CF_2 < \begin{matrix} CFCl \\ CFI \\ CCl_2 \end{matrix} \qquad [272]$$

One feature which can be attributed to the steric effect of fluorine is the preference of forming four-membered rings that is inherent to polyfluorinated olefins and which was not observed with other halogen derivatives **[273]**.

(136)

$$CHCl{=}CCl_2 \longrightarrow CHCl{=}CClCHClCCl_3$$

$$CFCl{=}CF_2 \longrightarrow \begin{matrix} CFCl{-}CF_2 \\ | \quad\quad | \\ CFCl{-}CF_2 \end{matrix} \qquad [273]$$

## IMPORTANT FEATURES IN THE REACTIVITY OF ORGANIC FLUORINE COMPOUNDS

Fluorine is in a way unique among the halogens, and deviates strongly from them in many respects. One of the differences is in bond energies of carbon–fluorine bonds, which are conspicuously stronger than those of the other halogens **[274]** (Table 37).

This difference shows in the rate of replacement of fluorine or chlorine, respectively, by other elements or groups. Both unimolecular and bimolecular mechanisms underly these substitution reactions. The ratios of F/Cl reactivity in various types of compounds and various reactions differ over a very wide range, from 0.1 to 0.00001, though the range 0.1–0.001 could be considered with greater justification. Replacement of fluorine is easy in compounds in which the $S_Ni$ mechanism is operating and where six- or five-membered rings can be closed to form products or intermediates **[275]** (Table 38).

Because of hydrogen–fluorine bonds (which have no analogs with other

**Table 37. Bond Energies of Fluoro Compounds Compared with Their Chloro Analogs [274]**

| | | | |
|---|---|---|---|
| $CH_3$–F | 107, 123 kcal/mole | $(C_6H_5)_3C$–F | 115 kcal/mole |
| $CH_3$–Cl | 81 kcal/mole | $(C_6H_5)_3C$–Cl | 86 kcal/mole |

**Table 38. Velocity Constants of Displacement of Fluorine by Carboxylate Ion at 65°C [275]**

| Fluoro compound | $F(CH_2)_5CO_2H$ | $F(CH_2)_4CO_2H$ | $F(CH_2)_3CO_2H$ |
|---|---|---|---|
| Intermediate | — | 6–membered lactone | 5–membered lactone |
| Probable mechanism | $S_N2$ | $S_Ni$ | $S_Ni$ |
| Velocity constant $k_1 \times 10^5$ sec$^{-1}$ | 0.006[a] | 1.37 | 34.75 |
| Relative reactivity | 1 | 230 | 5800 |

*a* Estimated.

halogens), hydrolytic displacement of fluorine is assisted by hydrogen ions, and acid catalysis was observed in quite a few examples.

Nucleophilic displacement of fluorine in systems like $-CF{=}CF-$, $-CF_2-$, and $C_6F_6$ shows a much higher reaction rate than displacement of other halogens in the corresponding systems.

(137) Nucleophilic displacement easier with

$$-CF{=}CF- \longrightarrow -CF{=}CY-$$
$$-CF_2- \qquad -CY_2-$$
$$C_6F_6 \qquad C_6F_5Y$$

than with the corresponding chloro derivatives

In addition to reactions in which fluorinated compounds differ from other halogen compounds by the rate of reaction, there are reactions that are unique for fluorine compounds and are not encountered among other halogen compounds, such as, for example, nucleophilic additions of alcohols, phenols, marcaptans, thiophenols, and amines to fluoro- and fluorohalo-olefins **[276,277]**.

(138) $$CF_2{=}CF_2 + ROH \xrightarrow{\text{base}} CHF_2CF_2OR$$ **[276]**

(139) $$CF_2{=}CF_2 + R_2NH \longrightarrow CHF_2CF_2NR_2$$ **[277]**

Another reaction unparalleled outside fluorine chemistry is the cleavage of perfluorinated ethers to acyl fluorides by aluminum chloride **[278]**.

(140) $$C_3F_7CF_2OCF_2C_3F_7 \xrightarrow{AlCl_3} C_3F_7COCl + C_3F_7CCl_3$$ **[278]**

The decarboxylation of perfluorocarboxylic acid salts to give monohydrylperfluoroparaffins or perfluoro-olefins is also peculiar to fluorine compounds **[279]**.

(141) $CF_3CF_2CF_2CHF_2 \xleftarrow[170\text{–}190°C]{(CH_2OH)_2} CF_3CF_2CF_2CF_2CO_2K \xrightarrow{165\text{–}200°C} CF_3CF_2CF{=}CF_2$ **[279]**

Some of the rearrangements of fluorinated compounds give products other than expected from the nonfluorinated compounds, e.g., Hofmann degradation of amides (p. 147). Other rearrangements are not exhibited at all by nonfluorinated compounds, such as, for example, rearrangements during the addition of fluorine to phenylethylene **[280,281]**.

(142) $$(C_6H_5)_2C{=}CH_2 \xrightarrow[Pb(OAc)_4]{HF} C_6H_5CF_2CH_2C_6H_5$$ **[280]**

(143) $$C_6H_5CH{=}CHC_6H_5 \xrightarrow{p\text{-}ClC_6H_4IF_2} (C_6H_5)_2CHCHF_2$$ **[281]**

By far the most important rearrangements of fluorinated compounds are shifts of fluorine to form polyfluoroalkyl clusters in fluorohaloethanes and propanes **[282]**.

(144) $$CF_2ClCCl_2F \xrightarrow{AlCl_3} CF_3CCl_3$$ **[282]**

## REDUCTION

Reduction methods consist of catalytic hydrogenation, complex hydride reduction, reduction with metals or metallic compounds, and reduction with organic compounds. Of these, the first two methods are of utmost importance. Catalytic hydrogenation is more suited for saturation of multiple bonds, the other methods of reduction more suitable for hydrogenolysis of single bonds (Table 39). Table 41 is a guide to practical applications of various reduction methods.

**Table 39. Relative Suitabilities of Reduction Methods**

| Method of reduction | Saturation of multiple bonds | | Hydrogenolysis of single bonds |
|---|---|---|---|
| | Isolated | Conjugated | |
| Catalytic hydrogenation | *** | ** | * |
| Complex-hydride reduction | — | ** | *** |
| Reduction with metals | — | * | ** |
| Reduction with metal salts | — | — | * |

## Catalytic Hydrogenation

The conditions of catalytic hydrogenation of fluorinated derivatives, especially the choice of catalysts, are similar to those for nonfluorinated compounds. Selective hydrogenation can be applied to systems which are capable of reduction to various degrees, as shown in Table 40 **[283]**.

Fluorine atoms usually resist hydrogenolysis in catalytic hydrogenation. However, there are several examples in which fluorine was replaced by hydrogen, sometimes under very mild conditions **[284]**.

$$(CF_3)_2C{=}CF_2 \xrightarrow[\text{H}_2\text{, Raney Ni, 100°C}]{\text{H}_2\text{, Pd, 20°C}} (CF_3)_2CHCHF_2 + (CF_3)_2CHCH_3 \qquad (145) \quad [284]$$

(H₂, Pd, 20°C: $(CF_3)_2CHCHF_2$ 95%; H₂, Raney Ni, 100°C: $(CF_3)_2CHCHF_2$ 10%, $(CF_3)_2CHCH_3$ 75%)

Fluorine bonded to an aromatic ring is replaced by hydrogen before the aromatic nucleus is hydrogenated. Fluorobenzoic acid first gives benzoic acid, and ultimately cyclohexanecarboxylic acid over platinum black **[285]**. Similarly, hydrogenolysis of fluorine precedes hydrogenation of the aromatic ring in *p*-fluorophenylacetic acid **[286]**.

**Table 40. Reduction of 4-Fluoro-3-nitroacetophenone under Various Conditions [283]**

| 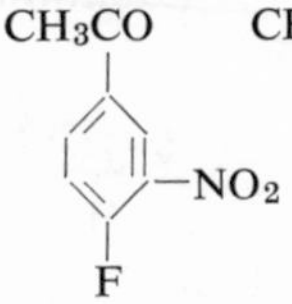 $CH_3CO$–C₆H₃(F)–$NO_2$ | 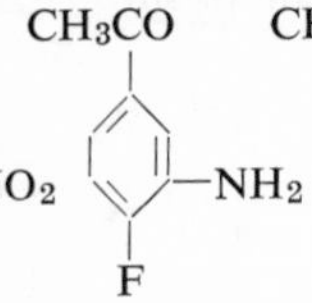 $CH_3CO$–C₆H₃(F)–$NH_2$ | 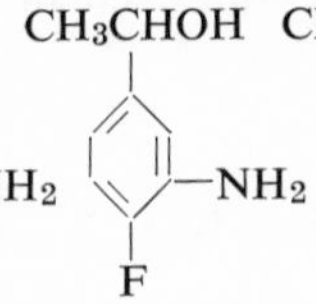 $CH_3CHOH$–C₆H₃(F)–$NH_2$ | 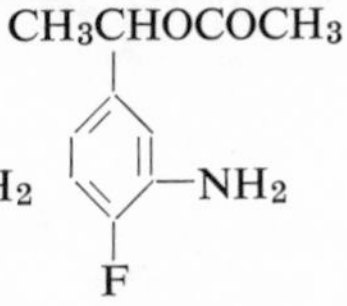 $CH_3CHOCOCH_3$–C₆H₃(F)–$NH_2$ | 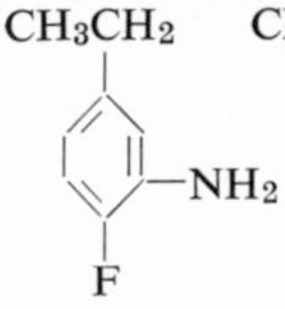 $CH_3CH_2$–C₆H₃(F)–$NH_2$ | 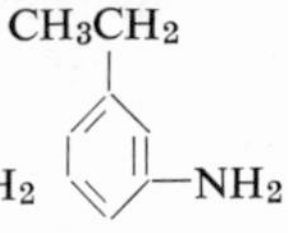 $CH_3CH_2$–C₆H₄–$NH_2$ |
|---|---|---|---|---|---|
| $3H_2$, Pd black, AcOH, $H_2SO_4$ | 60% | — | — | — | — |
| Pd black, AcOH | — | 35% | — | — | — |
| $4H_2$, Pd black, AcOH, $H_2SO_4$ | — | — | 41% | $H_2$, black, 60% | — |
| Pd black, AcOH, $Ac_2O$, $H_2SO_4$ | — | — | — | 63% | — |
| Pd(C), AcOH, $Ac_2O$, $H_2SO_4$ | — | — | — | 80% | — |
| Pd black, AcOH, $H_2SO_4$ | — | — | — | 80% | 20% |
| $NaBH_4$ | — | 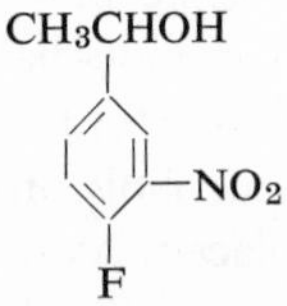 $CH_3CHOH$–C₆H₃(F)–$NO_2$ | — | — | — |

(146) [286]

$$F\text{-}C_6H_4\text{-}CH_2CO_2C_2H_5 \xleftarrow[200°C,\ 165\ atm,\ 2\ hr]{H_2,\ Raney\ Ni,\ EtOH} F\text{-}C_6H_4\text{-}CH_2CO_2H \xrightarrow[200°C,\ 160\ atm,\ 3.5\ hr]{H_2,\ Raney\ Ni,\ EtOH} C_6H_{11}\text{-}CH_2CO_2C_2H_5$$

However, the conventional and reliable method for hydrogenolysis of halogens over palladium on calcium carbonate in alkaline medium failed in the case of fluorobenzoic acid.

Fluorine in perfluorobenzene is replaced by hydrogen over platinum or palladium to give mainly pentafluorobenzene with small amounts of products containing two and three atoms of hydrogen **[287]**. Another way to pentafluorobenzene is desulfuration of pentafluorothiophenol **[288]**.

(147) [287, 288]

$$C_6F_6 \xrightarrow[40\text{–}60\%]{H_2,\ Pt\ or\ Pd,\ C} C_6HF_5 \xleftarrow[60\%]{Raney\ Ni,\ BuOH,\ reflux} C_6F_5SH$$

In the hydrogenation of pentafluorobenzene, the fluorine atom *para* to the hydrogen atom is replaced preferentially. Similarly, in pentafluoropyridine, mainly fluorine in position 4 (*para* to nitrogen) is replaced by hydrogen to give tetrafluoropyridine. In 3-chlorotetrafluoropyridine, catalytic hydrogenation preferentially replaces chlorine, whereas lithium aluminum hydride reduction preferentially replaces fluorine in position 4 [Equation (152)] **[289]**.

(148) [289]

$$C_5F_5N \xrightarrow[320°C]{H_2,\ Pd,\ C} \text{4-H-}C_5F_4N\ (30\%) + \text{2-H-}C_5F_4N\ (5\%) + \text{3-H-}C_5F_4N\ (trace)$$

## Reduction with Complex Hydrides

Reduction with fluorinated compounds with complex hydrides, especially with *lithium aluminum hydride*, competes successfully with catalytic hydrogenation as far as general applicability is concerned. It differs in the selectivity and types of compounds to which it is applied, and in this respect the two methods are complementary. The domain of complex-hydride reduction is the reduction of polar multiple bonds such as carbonyl or nitrile functions, and hydrogenolysis of carbon–halogen bonds, including carbon–fluorine bonds. It is especially suitable for the reduction of aldehydes, ketones, esters, acids, and their halides to alcohols, and of amides and nitriles to amines. Halogens are frequently replaced. As for fluorine, the carbon–fluorine bond in saturated chains resists hydrogenolysis **[290]**.

(149) [290]

$$\begin{array}{c}CF_2\text{—}CHF\\ |\quad\;\; |\\ CF_2\text{—}CFBr\end{array} \xrightarrow[63\%]{LiAlH_4,\ Et_2O} \begin{array}{c}CF_2\text{—}CHF\\ |\quad\;\; |\\ CF_2\text{—}CHF\end{array} \xleftarrow[51\%]{} \begin{array}{c}CF_2\text{—}CFCl\\ |\quad\;\; |\\ CF_2\text{—}CFI\end{array}$$

$$\begin{array}{c}CF_2\text{—}CFCl\\ |\quad\;\; |\\ CF_2\text{—}CFCl\end{array} \xrightarrow{74\%} \begin{array}{c}CF_2\text{—}CHF\\ |\quad\;\; |\\ CF_2\text{—}CHF\end{array} \xleftarrow{60\%} \begin{array}{c}CF_2\text{—}CFBr\\ |\quad\;\; |\\ CF_2\text{—}CFBr\end{array}$$

On the other hand, it is readily cleaved by complex hydrides in unsaturated compounds containing vinylic fluorine. Since the attacking species in lithium aluminum hydride reductions is the hydride ion, a carbon atom

(150)

$$\begin{array}{c}\text{—}CF_2\\ |\\ CCl\\ \|\\ \text{—}CF\end{array} \longleftrightarrow \begin{array}{c}\text{—}CF_2\\ |\\ :CCl\\ |\\ \text{—}\overset{\ominus}{C}F\end{array} \xrightarrow{\overset{\ominus}{H}:} \begin{array}{c}\text{—}CF_2\\ |\\ :\overset{\ominus}{C}Cl\\ |\\ \text{—}CHF\end{array} \xrightarrow{-\overset{\ominus}{F}} \begin{array}{c}\text{—}CF_2\\ |\\ CCl\\ \|\\ \text{—}CH\end{array} \;+\; \begin{array}{c}\text{—}CF\\ \|\\ CCl\\ |\\ \text{—}CHF\end{array}$$

(151) [291]

LiAlH$_4$ + F—C=C—F ring (C=C, $CF_2$, $CF_2$, $CF_2$)

$S_N2$ → H—C=C—F ($CF_2$, $CF_2$, $CF_2$)

$S_N2'$ → H—CF—C(F)=CF ($CF_2$, $CF_2$)

From H—C=C—F: $S_N2$ → H—C=C—H ($CF_2$, $CF_2$, $CF_2$); $S_N2'$ → H—C(=CF)—CF(H) ($CF_2$, $CF_2$)

From H—CF—C(F)=CF: $S_N2$ → H—C(=CF)—CF(H) ($CF_2$, $CF_2$); $S_N2$ → H—CF—C(F)=CH ($CF_2$, $CF_2$)

From H—C=C—H: $S_N2'$ → H—CH—C(H)=CF ($CF_2$, $CF_2$)

From H—C(=CF)—CF(H) ($S_N2$) and H—CF—C(F)=CH ($S_N2$) → H—C(=CH)—CF(H) ($CF_2$, $CF_2$)

From H—C(=CH)—CF(H): $S_N2$, $S_N2'$ → H—C(=CH)—CH(H) ($CF_2$, $CF_2$); $S_N2'$ → H—CH—CF(H) (CH=CF, $CF_2$)

carrying fluorine is attacked preferentially to a carbon atom carrying chlorine or other halogens (because of the stronger mesomeric effect of fluorine). Consequently, in vicinal fluorohalo-olefins, fluorine is usually replaced, either by the $S_N2$ or $S_N2'$ reaction mechanism **[291]**.

In the aromatic series, fluorine is replaced by hydrogen preferentially to chlorine in the reaction of polyfluorochlorocompounds with lithium aluminum hydride (in contrast to catalytic hydrogenation, which replaces chlorine preferentially) **[292]**. Nevertheless, even chlorine is displaced by hydrogen, so that a complex mixture of products results from the reduction of chloropentafluorobenzene with lithium aluminum hydride **[293,294]**.

(152) [292]

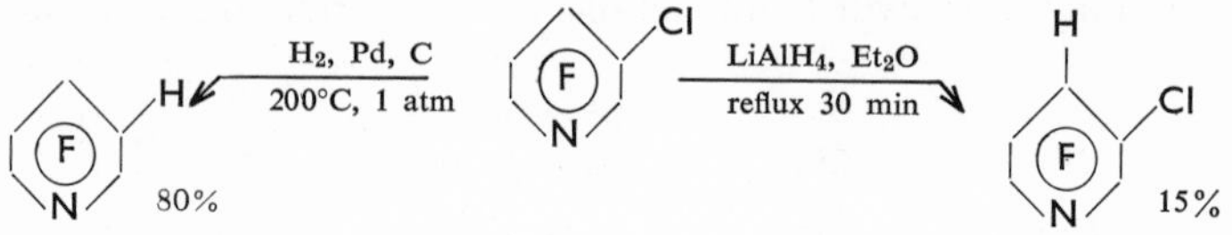

$H_2$, Pd, C; 200°C, 1 atm → 80%

$LiAlH_4$, $Et_2O$; reflux 30 min → 15%

(153) [293, 294]

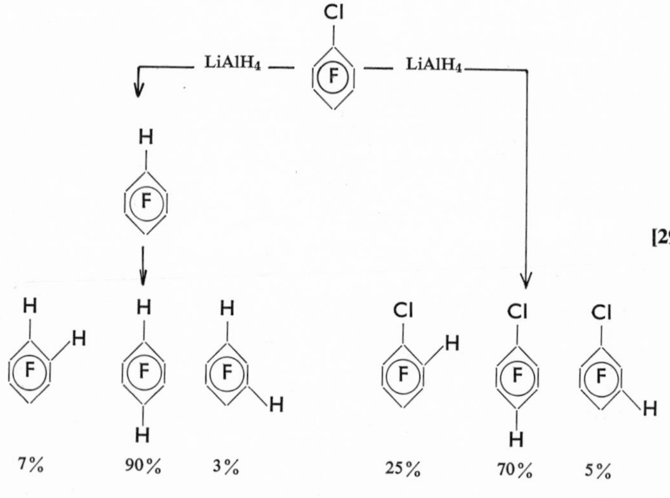

$LiAlH_4$ — $LiAlH_4$

7% 90% 3% 25% 70% 5%

In perfluoroalkylisocyanates and N-perfluoroalkylcarbamates, the two fluorine atoms adjacent to the nitrogen atom are replaced by hydrogen by means of lithium aluminum hydride **[295,296]**.

(154) [295, 296]

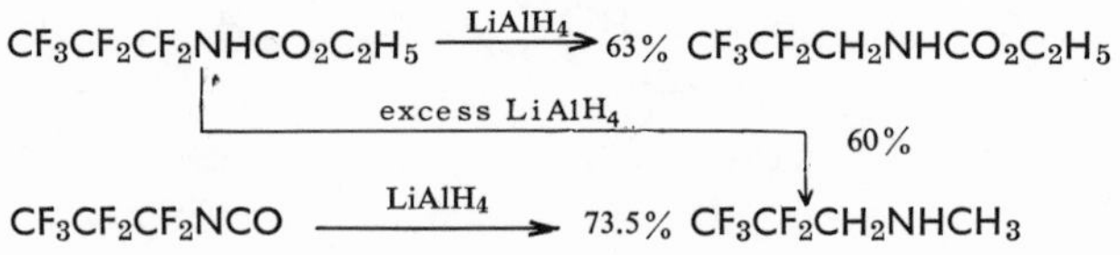

$$CF_3CF_2CF_2NHCO_2C_2H_5 \xrightarrow{LiAlH_4} 63\%\ CF_3CF_2CH_2NHCO_2C_2H_5$$

excess $LiAlH_4$ 60%

$$CF_3CF_2CF_2NCO \xrightarrow{LiAlH_4} 73.5\%\ CF_3CF_2CH_2NHCH_3$$

## Reduction with Metals and Metallic Compounds

Metals are frequently used for hydrogenolysis of halogens, including fluorine. Reductions with *sodium* usually result in the replacement of fluorine by hydrogen **[297,298]**.

(155) $C_6H_5CF{=}CFC_6H_5 \xrightarrow{\text{Na, MeOH}} C_6H_5CH_2CH_2C_6H_5$ 81% **[297]**

(156) $C_6H_5COCF_2CF_2CO_2H \xrightarrow[C_7H_8,\ \text{reflux}]{\text{Zn, HgCl}_2,\ \text{HCl}} C_6H_5CH_2CH_2CH_2CO_2H$ **[298]**

*Zinc* usually does not attack carbon–fluorine bond, whereas other halogens are readily replaced by hydrogen. In the case of vicinal fluorohalides, zinc in protic solvents substitutes hydrogen for the halogen, and in aprotic solvents eliminates both to form an olefin **[299]**.

(157) **[299]**

$CF_3CHClBr \xrightarrow[\text{reflux}]{\text{Zn, MeOH}} CF_3CH_2Cl$ 58.7%

$CF_3CHClBr \xrightarrow[\text{reflux}]{\text{Zn, dioxane}} CF_2{=}CHCl$ 71%

*Iron* is suitable for partial reduction of polyhalogen clusters **[300]**. The same kind of reduction can also be achieved by *sodium sulfite* **[301]**.

(158)

$CF_3CClBr_2 \xrightarrow[64°C]{\text{Fe, HCl}} CF_3CHClBr$ 45% **[300]**

$CF_3CClBr_2 \xrightarrow[65\text{–}72°C]{Na_2SO_3,\ \text{NaOH}} CF_3CHClBr$ 80–90% **[301]**

*Tin* and *stannous chloride* reduce nitro and azo groups to amino groups and diazo compounds to hydrazines. *Hydrogen iodide* is suitable for the replacement of an alcoholic hydroxyl group by hydrogen.

## Reduction with Organic Compounds

The only practical example of the reduction of fluorinated compounds by organic reagents is the *Meerwein–Ponndorf reduction* of aldehydes and ketones, which is of advantage because of its selectivity.

A few instances of the reduction of aldehydes, ketones, esters, and nitriles with Grignard reagents, especially those derived from secondary and tertiary bromides or iodides, show the danger of side-reactions in the Grignard synthesis but are hardly of practical importance (p. 120).

A brief survey of various methods of reduction is given in Table 41.

**Table 41. Selectivity of Reducing Reagents[a]**

| Bond or function | C=C | C≡C | Aromatic system | C=C conjug. | CO | COCl, $CO_2R$ | $CO_2H$, $CONR_2$ | $CONH_2$ | C≡N | C—Hal | C—F | =CF | $NO_2$ | N=N | N≡N | $SO_2Cl$ |
|---|---|---|---|---|---|---|---|---|---|---|---|---|---|---|---|---|
| $H_2$ Catal. | ** | ** | ** | — | ** | ** | — | ** | ** | ** | * | * | ** | — | — | — |
| $LiAlH_4$ | — | — | — | * | ** | ** | ** | ** | ** | ** | * | ** | — | — | — | — |
| Na | — | — | — | * | — | — | — | — | — | — | * | * | — | — | — | — |
| Zn | — | — | — | — | — | — | — | — | — | ** | * | † | — | — | — | ** |
| Fe | — | — | — | — | — | — | — | — | — | ** | — | — | ** | — | — | — |
| Sn | — | — | — | — | — | — | — | — | — | — | — | — | ** | ** | — | — |
| $SnCl_2$ | — | — | — | — | — | — | — | — | — | — | — | — | — | — | ** | — |
| Alcohols | — | — | — | — | * | — | — | — | — | — | — | — | — | — | — | — |
| Grignard reagents | — | — | — | — | * | — | * | — | * | — | — | — | — | — | — | — |

[a] (*) Applicable; (**) generally used; (†) unsuitable.

## OXIDATION

In industry, oxidations are usually carried out with oxygen, whereas in the laboratory, sodium or potassium dichromate and potassium permanganate are the most common oxidizing agents. Only relatively few reactions are carried out with more selective oxidation reagents such as hydrogen peroxide, nitric acid, manganese dioxide, mercuric oxide, lead tetraacetate, or halogens.

### Oxidations with Oxygen

The most deeply explored oxidation by oxygen is that leading to trifluoroacetic acid from trifluoroethane **[302]** or its chloro derivatives **[303]**, or from olefins containing trifluoromethyl group **[304]**.

(159)

$$CF_3CH_3 \xrightarrow[15000\ V]{O_2,\ H_2O} CF_3CO_2H\ (49\%)\ [302]$$

$$CF_3CH_2Cl \xrightarrow{Cl_2,\ O_2,\ 200°C} CF_3CO_2H\ (72\%)\ [303]$$

$$CF_3CHCl_2 \xrightarrow{Cl_2,\ O_2,\ 200°C} CF_3CO_2H\ (93\%)\ [303]$$

(160)

$$CF_3CCl{=}CClCF_3 \xrightarrow{O_2} \underset{83\%}{CF_3COCl} + (CF_3CO_2H) \qquad [304]$$

Much attention has been devoted to the oxidation of tetrafluoroethylene and chlorotrifluoroethylene to the corresponding epoxides, which on rearrangement give fluorides of trifluoroacetic or chlorodifluoroacetic acid, respectively **[305,306]**.

(161)

$$CClF{=}CF_2 \xrightarrow[-80°C]{O_2} \underset{25\%}{CClF{-}CF_2\ (\text{epoxide, }{-}O{-})}\ [305]$$

$$CClF{=}CF_2 \xrightarrow[25\text{–}50°C,\ 7\text{–}21\ atm]{O_2} \underset{43\%}{CClF_2COF}\ [306]$$

### Oxidations with Oxidative Reagents

The majority of oxidations are directed toward the preparation of aldehydes, ketones, and, most commonly, acids starting from fluorinated olefins.

*Oxidations of Fluoro-olefins*

A nondestructive oxidation of perfluoro-2-butene, 1,1,1,3,3,3,-hexafluoro-2-butene, 2,3-dichloro-1,1,1,3,3,3-hexafluorobutene, and perfluoro-2-butyne with *chromic acid* gives hexafluorodiacetyl **[307]**.

(162) [307]

$$\begin{array}{ccccc} CF_3CH{=}CHCF_3 & \xrightarrow{26.6\%} & & \xleftarrow{35\%} & CF_3CCl{=}CClCF_3 \\ & & CF_3COCOCF_3 & & \\ CF_3C{\equiv}CCF_3 & \xrightarrow{17.1\%} & & \xleftarrow{7.3\%} & CF_3CF{=}CFCF_3 \end{array}$$

$CrO_3$, $H_2SO_4$, $SO_3$, 60–70°C

Degradative oxidation of fluorinated olefins with *potassium permanganate* is the best laboratory procedure for the preparation of fluorinated ketones and acids **[308–312]**.

(163)

$$(CF_3)_2C{=}CF_2 \xrightarrow[\text{[309]}]{1.\ KMnO_4\quad 2.\ P_2O_5} 31\%\ CF_3COCF_3$$

$$(CF_3)_2C{=}CH_2 \xrightarrow{[308]} 16\%\ CF_3COCF_3\ 55\% \xleftarrow{[308]} (CF_3)_2C{=}CHBr$$

(164)

$$CF_3CCl{=}CClCF_3 \xrightarrow[\text{[311]}]{KMnO_4\ \text{reflux}} 83\%\ CF_3CO_2H$$

$$CF_3CH{=}CH_2 \xrightarrow[80°C]{KMnO_4} 80\%\ CF_3CO_2H\ 90\% \xleftarrow[65–70°C]{KMnO_4} CF_3CCl{=}CCl_2$$

[310] [312]

Under the same conditions, cyclic fluoro-olefins give fluorinated dicarboxylic acids. The reaction is best carried out at low temperature in aqueous acetone solution **[313]**.

(165) cyclo-$(CF_2)_4CF{=}CF$ $\xrightarrow{KMnO_4,\ (CH_3)_2CO,\ 20°C}$ $HO_2C(CF_2)_4CO_2H$ 75% **[313]**

(166) cyclo-$CF{=}CF{-}CF_2{-}CF{=}CF{-}CF_2$ $\xrightarrow{KMnO_4,\ (CH_3)_2CO,\ 20°C}$ $HO_2CCF_2CO_2H$ 60% **[313]**

*Oxidation of Fluorinated Aromatics*

Fluorinated aromatic hydrocarbons having side chains and fluorine atoms in the nucleus are frequently oxidized to fluorinated aromatic aldehydes or acids. For fluorinated benzaldehydes, halogenation to the stage of benzal halides followed by hydrolysis, or halogenation to the stage of benzyl halides followed by treatment with *lead or copper nitrate*, gives approximately the same yield, superior to that reached by chromyl chloride oxidation of fluorotoluenes **[314,315]**.

(167) $FC_6H_4CH_3 \xrightarrow{Br_2,\ h\nu,\ 110°C} FC_6H_4CH_2Br \xrightarrow{Pb(NO_3)_2,\ reflux} FC_6H_4CHO$ 60 – 70% **[314]**

$FC_6H_4CH_3 \xrightarrow{CrO_2Cl_2,\ CCl_4} FC_6H_4CHO$ 50% **[315]**

$FC_6H_4CH_3 \xrightarrow{Cl_2,\ h\nu} FC_6H_4CHCl_2 \xrightarrow{CaCO_3,\ reflux} FC_6H_4CHO$ 60–70% **[314]**

Oxidation of ring-fluorinated benzene homologs with dilute *nitric acid* gives good yields of fluorobenzoic acids. The reaction requires stainless steel autoclaves **[316]**.

(168) $$FC_6H_4CH_3 \xrightarrow[190\text{–}195°C]{20\%\ HNO_3} FC_6H_4CO_2H \quad 82\text{–}96\%$$ **[316]**

Oxidative degradation of an aromatic ring to a carboxylic group takes place in the energetic oxidation of *m*-aminobenzotrifluoride with sodium dichromate **[317]** or potassium permanganate **[318]**. This first preparation of trifluoroacetic acid remains the easiest way to this compound if trifluoromethyl-group-containing olefins are not available.

(169) $m\text{-}CF_3C_6H_4NH_2 \xrightarrow[70\text{–}170°C]{Na_2Cr_2O_7,\ H_2SO_4} CF_3CO_2H$ 90–95% **[317]**

$m\text{-}CF_3C_6H_4NH_2 \xrightarrow[80°C,\ reflux]{KMnO_4,\ H_2O,\ AcOH} CF_3CO_2H$ 73.5% **[318]**

*Oxidation of Nitrogen and Sulfur Compounds*

Fluorinated ketone hydrazones are converted to diazo compounds by *mercuric oxide or lead tetraacetate* **[319]**, and fluorinated sulfides are oxidized to sulfoxides or sulfones, depending on the reagent and conditions **[320]**. Fluorinated thiocyanates and isothiuronium salts are oxidized with *chlorine* to fluorinated sulfonyl chlorides **[321,322]**.

(170) $$(CF_3)_2C{=}NNH_2 \xrightarrow[25°C,\ 1\ hr]{Pb(OAc)_4,\ PhCN} (CF_3)_2C{=}N{=}N \quad 77\%$$ **[319]**

(171) $$(FCH_2CH_2)_2SO \xleftarrow{HNO_3} (FCH_2CH_2)_2S \xrightarrow{K_2Cr_2O_7} (FCH_2CH_2)_2SO_2 \quad 65\%$$ **[320]**

**Table 42. Selectivity of Oxidating Reagents**

| Oxidation reagent | $CH_3 \rightarrow CO_2H$ | $C{=}C \rightarrow CO_2H$ | $CH_2OH \rightarrow CHO$ | $CH_2OH \rightarrow CO_2H$ | Arom. $\rightarrow CO_2H$ | $NNH_2 \rightarrow N_2$ | $NO \rightarrow NO_2$, $NH_2 \rightarrow$ | $S \rightarrow SO, SO_2$, $\rightarrow SO_3H$ |
|---|---|---|---|---|---|---|---|---|
| $O_2$ | * | * | — | — | — | — | — | — |
| $H_2O_2$ | — | — | — | — | — | — | * | * |
| $HNO_3$ | * | — | — | — | — | — | — | * |
| $Cl_2$, $H_2O$ | — | — | * | — | — | — | — | * |
| $HgO$ | — | — | — | — | — | * | — | — |
| $MnO_2$ | — | — | * | — | — | — | — | — |
| $Pb(OAc)_4$ | — | — | — | — | — | * | — | — |
| $Na_2Cr_2O_7$ | — | — | — | * | * | — | — | * |
| $KMnO_4$ | * | * | — | * | * | — | — | — |

(172) $FCH_2CH_2SCN$ $\xrightarrow[{[322]}]{Cl_2,\ H_2O}$ 60.6% $FCH_2CH_2SO_2Cl$ 51% $\xleftarrow[{[321]}]{Cl_2,\ H_2O}$ $FCH_2CH_2SC(=NH)NH_2 \cdot TosOH$

*Anodic Oxidation*

The electrolysis of fluorinated alkanecarboxylic acids in methanol in the presence of sodium methoxide leads to anodic decarboxylation and coupling to give fluorinated alkanes. In this way, different acids can also be combined to form nonsymmetrical products **[323]**.

$$F(CH_2)_nCO_2H \xrightarrow[110\ V,\ 1\text{–}2\ A]{MeONa} F(CH_2)_{2n}F \qquad \begin{matrix} n: & 4 & 5 & 6 & 9 \\ \%: & 45 & 45 & 58 & 69 \end{matrix} \qquad [323] \tag{173}$$

A guide to different types of oxidations and suitable oxidation reagents is given in Table 42.

## ELECTROPHILIC REACTIONS

While this section will include mainly electrophilic reactions (additions and substitutions), reactions with other mechanisms will also be discussed when similarities or some common features can be pointed out.

### Halogenation

In addition to proper halogenation, reactions leading to halogen derivatives in general will be mentioned, such as reactions of organic fluoro compounds with hydrogen halides, and nonmetal and metal halides.

*Addition of Halogens across Multiple Bonds*

Addition of halogens to olefins is a typical *trans* addition, although mixtures of *cis* and *trans* products were sometimes obtained. Chlorine and 1H,2H-octafluorocyclohexene give DL-1H,2H,1,2-dichlorooctafluorocyclohexane with both chlorines in equatorial positions **[324]**.

(174) [structure: 1H,2H-octafluorocyclohexene] $\xrightarrow{Cl_2}$ [structure: 1H,2H-1,2-dichlorooctafluorocyclohexane, both Cl equatorial] **[324]**

Addition of iodine to tetrafluoroethylene requires elevated temperature, which is rather surprising since vicinal diiodides usually split out iodine when heated **[325–327]**.

(175)
$$CF_2{=}CF_2 \xrightarrow[\text{60°C, 21–25 atm}]{I_2,\ Et_2O} CF_2ICF_2I\ (74\%) \quad [325]$$
$$CF_2{=}CF_2 \xrightarrow[\text{autoclave}]{I_2,\ 150°C} CF_2ICF_2I\ (76\%) \quad [326]$$
$$CF_2{=}CF_2 \xrightarrow[\text{100°C}]{I_2,\ KI,\ H_2O} CF_2ICF_2I\ (91\%) \quad [327]$$

The addition of interhalogen compounds to nonsymetrical fluoroolefins is sometimes bidirectional **[328,329]**.

(176) **[328]**
$$CF_2{=}CFCl \xrightarrow{ICl} CF_2ICFCl_2\ (29\%) + CF_2ClCFClI\ (57\%)$$
$$CF_2{=}CFCl \xrightarrow{IBr} CF_2ICFClBr\ (14\%) + CF_2BrCFClI\ (44\%)$$

*Replacement of Hydrogen by Halogens*

The halogenation of fluorinated paraffins is a free-radical reaction, and consequently occurs preferentially at carbon atoms that are prone to split off hydrogen atoms homolytically. Therefore, hydrogen atoms on a carbon atom adjacent to a difluoromethylene or trifluoromethyl group are, as a rule, not replaced because their bonds are too polar **[330]**.

(177) **[330]**

$CF_3CH_2CH_2CH_3$ $\xrightarrow{Cl_2,\ h\nu \mid H_2O}$ $CF_3CH_2CHClCH_3$ (4p.) + $CF_3CH_2CH_2CH_2Cl$ (5p.)

$CF_3CH_2CHClCH_3$ → $CF_3CH_2CCl_2CH_3$ (8p.) + $CF_3CH_2CHClCH_2Cl$ (6p.)

$CF_3CH_2CH_2CH_2Cl$ → $CF_3CH_2CHClCH_2Cl$ (1p.) + $CF_3CH_2CH_2CHCl_2$ (2p.)

$CF_3CH_2CH_2CHCl_2$ → $CF_3CH_2CH_2CCl_3$

**Table 43. Chlorination and Bromination of Fluorobenzene [331]**

| | Composition of $C_6H_4FCl$ | | | Composition of $C_6H_4FBr$ | | |
|---|---|---|---|---|---|---|
| Temperature, °C | %*o* | %*m* | %*p* | %*o* | %*m* | %*p* |
| 260 | 1.7 | 32.0 | 66.3 | 1.8 | 6.4 | 91.8 |
| 345 | 2.7 | 47.2 | 50.1 | 1.9 | 5.6 | 92.5 |
| 600 | 11.5 | 56.5 | 32.0 | 11.5 | 60.5 | 28.0 |

Also, the halogenation of side chains in aromatic fluoro compounds is a free-radical reaction and proceeds best under irradiation with ultraviolet light **[314]**. A free-radical mechanism is probably also operating in the high-temperature halogenation of fluorobenzene, judging from the high proportion of *meta* substitution at the expense of *ortho*/*para* substitution (Table 43) **[331]**.

On the other hand, the halogenation of fluorinated aromatics at mild temperatures in the presence of catalysts follows the rules for electrophilic substitution, i.e., the halogen enters *para* and *ortho* positions to an electron-releasing group, and *meta* position to an electron-withdrawing group. In *o*-fluorotoluene, halogenation occurs mainly in the *para* position to fluorine **[332]**, in benzotrifluoride in the *meta* position to the trifluoromethyl group **[333]**. An indirect method is used for introducing bromine into the *ortho* position to the trifluoromethyl group **[334]**.

(178) o-fluorotoluene ($CH_3$, F) —$Cl_2$, Fe→ 4-chloro-2-fluorotoluene ($CH_3$, F, Cl) 58.5% + 2-chloro-6-fluorotoluene (Cl, $CH_3$, F) 17.8% **[332]**

(179) $CF_3$-benzene —$Br_2$, Fe, 60–70°C→ m-bromobenzotrifluoride ($CF_3$, Br) 52% [333]; $CF_3$-benzene —BuLi, $Et_2O$→ o-lithiobenzotrifluoride ($CF_3$, Li) —$Br_2$→ o-bromobenzotrifluoride ($CF_3$, Br) 28% [334]

Strongly fluorinated aromatics require forcing conditions, but nevertheless give high yields of halogenation products **[335]**.

(180) $C_6HF_5$ —$Br_2$, $AlBr_3$, 20% oleum, 60–65°C→ $C_6F_5Br$ 80.8%; $C_6HF_5$ —I, 20% oleum→ $C_6F_5I$ 71% **[335]**

*Cleavage of the Carbon Chain by Halogens*

The halogenation of silver salts of perfluorinated carboxylic acids is accompanied by the elimination of carbon dioxide and results in the formation of perfluoroalkyl halides having one less carbon atom than the starting acid (Hunsdieckers' method) **[336]**.

(181) [336]

$$C_5F_{11}Cl \ 71.2\% \xleftarrow[100°C]{Cl_2} C_5F_{11}CO_2Ag \xrightarrow[80\text{–}90°C]{Br_2} C_5F_{11}Br \ 82.5\%$$

$$C_5F_{11}CO_2Ag \xrightarrow[100°C]{I_2} C_5F_{11}I \ 73.9\%$$

The high-temperaure chlorination and bromination of perfluoro-*p*-cymene cleaves the carbon chain and gives 4-halogenoperfluorotoluene and perfluoroisopropyl halides **[337]**.

(182) [337]

$$p\text{-}CF_3C_6F_{10}Cl \ (30.7\%) + CFCl(CF_3)_2 \xleftarrow{Cl_2,\ 540\text{–}555°C} p\text{-}CF_3C_6F_{10}CF(CF_3)_2 \xrightarrow{Br_2,\ 510°C} CFBr(CF_3)_2 \ (37\%) + p\text{-}CF_3C_6F_{10}Br \ (21\%)$$

*Addition of Hydrogen Halides across Multiple Bonds*

The direction of addition of hydrogen halides to fluorinated olefins is determinated by the polarity of the double bond, which in turn depends on the position of fluorine atoms in the molecule. In 1,1,1-trifluoropropene, halogen joins the terminal carbon atom. The same orientation takes place in the addition of hydrogen halides to perfluoropropene. Rather energetic conditions are necessary for this reaction **[338]**. The addition of hydrogen bromide to chlorotrifluoroethylene also requires considerable activation, and takes place in such a way that bromine joins the difluoromethylene end of the double bond both at polar and free-radical conditions **[339–342]**. The same orientation occurs in the addition of hydrogen halides to trifluoropropyne **[343]**.

(183) [338]

$$CF_3CF{=}CF_2 \longleftrightarrow CF_3\overset{\ominus}{\underset{\cdot\cdot}{C}}F{-}\overset{\oplus}{C}F_2$$

$$\xrightarrow{HF,\ CaSO_4,\ 200°C} CF_3CHFCF_3 \ 80\%$$

$$\xrightarrow{HCl,\ 230°C} CF_3CHFCF_2Cl \ 60\%$$

$$\xrightarrow{HBr} CF_3CHFCF_2Br \ 50\%$$

(184) 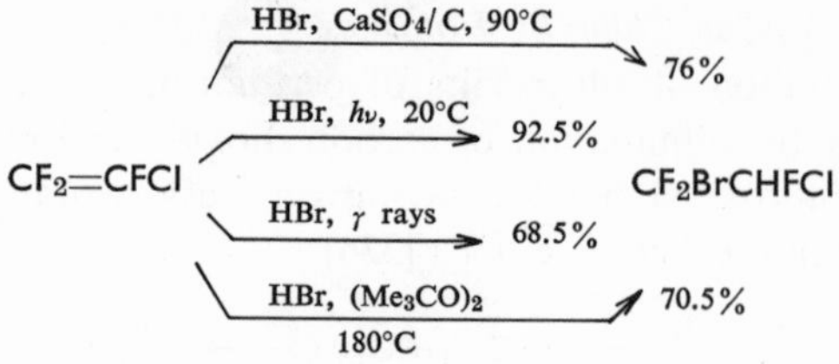 $CF_2{=}CFCl$ → $CF_2BrCHFCl$

- HBr, $CaSO_4$/C, 90°C → 76% [339]
- HBr, $h\nu$, 20°C → 92.5% [340]
- HBr, $\gamma$ rays → 68.5% [341]
- HBr, $(Me_3CO)_2$, 180°C → 70.5% [342]

(185) 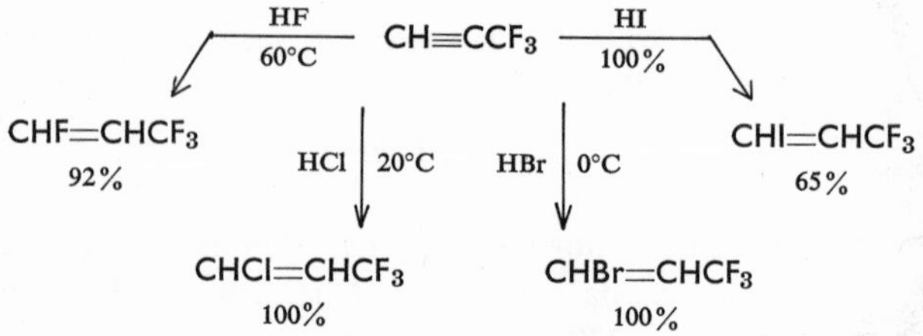 $CH{\equiv}CCF_3$ [343]

- HF, 60°C → $CHF{=}CHCF_3$ 92%
- HI, 100% → $CHI{=}CHCF_3$ 65%
- HCl, 20°C → $CHCl{=}CHCF_3$ 100%
- HBr, 0°C → $CHBr{=}CHCF_3$ 100%

*Replacement of Oxygen- and Nitrogen-Containing Functions by Halogens*

The replacement of a hydroxylic group in fluorinated alcohols can be achieved by hydrogen halides, but thionyl chloride and phosphorus halides give better results. These reagents can also be used for converting fluorinated acids to acyl halides. A favorite way for the replacement of a hydroxylic group by a halogen is the conversion of an alcohol to its tosyl ester, and treatment of the tosylate with an alkaline halide.

Fluorinated aliphatic diazo compounds react with hydrogen halides to form fluoroalkyl halides **[344]**. Fluorinated aromatic diazonium salts are converted to aromatic halides by the *Sandmeyer and Gattermann reactions.*

(186) 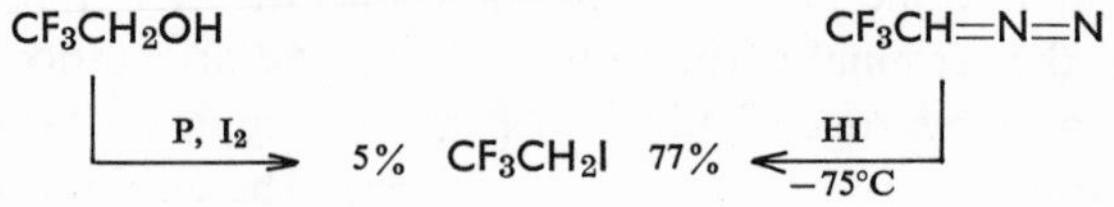 [344]

$$CF_3CH_2OH \xrightarrow{P,\ I_2} 5\%\ CF_3CH_2I\ 77\% \xleftarrow[-75^\circ C]{HI} CF_3CH{=}N{=}N$$

*Replacement of Fluorine by Other Halogens*

Some perfluoro-olefins are converted to perhalo-olefins under surprisingly mild conditions when treated with aluminum halides. All of the fluorine atoms are usually replaced **[345]**.

(187) 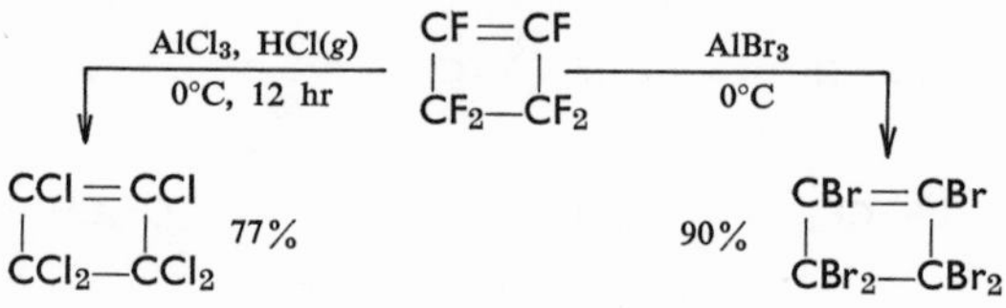 [345]

Cyclobutene $CF{=}CF$ / $CF_2{-}CF_2$:

- $AlCl_3$, HCl(g), 0°C, 12 hr → $CCl{=}CCl$ / $CCl_2{-}CCl_2$ 77%
- $AlBr_3$, 0°C → $CBr{=}CBr$ / $CBr_2{-}CBr_2$ 90%

In perfluoro-ethers, aluminum chloride replaces all fluorine atoms adjacent to the oxygen atom. In addition, the ethers that do not have side chains at carbon atoms neighboring to oxygen are cleaved to trichloroperfluoroacyl chlorides **[346]**.

(188) Perfluoro-2-butyltetrahydrofuran ($CF_2$—$CF_2$, $CF_2$, $CFC_4F_9$, O ring) $\xrightarrow[200–220°C]{AlCl_3}$ 2,2,5-trichloro-5-nonafluorobutyl-3,3,4,4-tetrafluorotetrahydrofuran ($CF_2$—$CF_2$, $CCl_2$, $CClC_4F_9$, O ring) 51% **[346]**

(189) Perfluorotetrahydropyran ($CF_2$, $CF_2$, $CF_2$, $CF_2$, $CF_2$, O ring) $\xrightarrow[180°C]{AlCl_3}$ $CCl_3CF_2CF_2CF_2COCl$ 38% **[346]**

## Nitration

Treatment of fluorinated olefins with *dinitrogen tetroxide* results in the formation of vicinal nitro-nitrites or dinitrites. The former are ultimately hydrolyzed to nitro-alcohols, the latter to hydroxy acids **[347]**.

(190) $(CF_3)_2C{=}CF_2 \xrightarrow[170–180°C]{N_2O_4} (CF_3)_2C(ONO)CF_2ONO + (CF_3)_2C(ONO)CF_2NO_2$ **[347]**

$(CF_3)_2C(ONO)CF_2ONO \rightarrow (CF_3)_2C(OH)CO_2H$ 35%

$(CF_3)_2C(ONO)CF_2NO_2 \rightarrow (CF_3)_2C(OH)CF_2NO_2$ 94%

Dinitrogen tetroxide reacts with trifluoroacetic acid to give trifluoronitromethane **[348]**. The same compounds are also obtained by treatment of trifluoroacetic anhydride with fuming nitric acid **[349]**.

(191) $CF_3CO_2H \xrightarrow[200°C]{N_2O_4} CF_3NO_2$ 30% **[348]**; $(CF_3CO)_2O \xrightarrow[100°C]{HNO_3(100\%)} CF_3NO_2$ 7% **[349]**

Nitration of fluorinated aromatics is carried out in conventional ways. Fluorobenzene is smoothly converted to *p*-nitrofluorobenzene, or 2,4-dinitrofluorobenzene **[350]**. Somewhat more energetic conditions are required for the nitration of pentafluorobenzene **[351,352]**.

(192) $C_6H_5F \xrightarrow[0–20°C]{HNO_3(100\%),\ H_2SO_4}$ 2,4-$(NO_2)_2C_6H_3F$ 87% **[350]**

(193)

$$C_6HF_5 \xrightarrow[60\text{–}70°C,\ 2\ hr]{HNO_3(95\%),\ BF_3,\ C_4H_8SO_2} C_6F_5NO_2 \quad 82.5\% \quad [351]$$

$$C_6HF_5 \xrightarrow[20°C]{HNO_3(100\%),\ HF} C_6F_5NO_2 \quad 50\% \quad [352]$$

## Nitrosation

Nitrosation at carbon was achieved in the preparation of trifluoronitrosomethane, which was obtained either by treatment of iodotrifluoromethane with nitric oxide **[353]**, or by decomposition of trifluoroacetyl nitrite resulting from the action of nitrosyl chloride upon silver trifluoroacetate **[354]**.

(194)

$$CF_3I\ [353] \xrightarrow[h\nu,\ 20°C]{NO,\ Hg} CF_3NO \quad 89\%$$

$$CF_3CO_2Ag\ [354] \xrightarrow[-20°C,\ 2\ hr]{NOCl} CF_3CO_2NO\ (92.5\%) \xrightarrow[3.5\ hr]{190\text{–}192°C} CF_3NO \quad 56\%$$

Nitrosation at nitrogen—*diazotization* in the case of primary amines—is important for the synthesis of aliphatic fluorinated diazo compounds **[355]**, and aromatic diazonium compounds. The latter give common coupling reactions **[356]**.

(195)

$$CF_3CH_2NH_2{\cdot}HCl \xrightarrow{NaNO_2,\ H_2SO_4} CF_3CH{=}N{=}N \quad 50\text{–}70\% \quad [355]$$

(196)

$$C_6F_5NH_2 \xrightarrow{HNO_2} C_6F_5N_2X \xrightarrow{C_6F_5NH_2} C_6F_5N{=}NNHC_6F_5 \quad [356]$$

## Sulfonation

Examples of sulfonation of fluorinated compounds are rather scarce. One of the most interesting achievements in this respect is sulfonation of pentafluorobenzene, which—surprisingly—does not require too strong conditions **[357]**.

(197)

$$C_6HF_5 \xrightarrow[15°C,\ 48\ hr]{20\%\ oleum} C_6F_5SO_3H \quad 88\% \quad [357]$$

## Acid-Catalyzed Syntheses (Friedel-Crafts Reaction)

A common feature of acid-catalyzed syntheses is the formation of an intermediate carbonium ion. In the aliphatic series, this carbonium ion adds across a double bond, and is followed by the addition of a negatively charged species to the other carbon of the double bond. In the aromatic series, the addition of the carbonium ion is followed by the elimination of a proton and regeneration of the aromatic system. Both reactions, addition and substitution, are catalyzed by Lewis acids.

*Acid-Catalyzed Additions*

Chloroparaffins and chlorofluoroparaffins add to fluorinated olefins and form homologous chlorofluoroparaffins. Both carbon–chlorine and carbon–fluorine bonds in the starting chlorofluoroparaffin are cleaved, carbon–fluorine bonds preferentially. Addition across the double bond occurs bidirectionally with nonsymmetrical olefins. Consequently, complex mixtures of products result, and the reaction does not have much practical importance for synthetic purposes **[358]**.

(198) **[358]**

$$CCl_3F \xrightarrow[21\text{–}23°C]{CCl_2{=}CF_2,\ AlCl_3} CF_3CCl_2CCl_3\ (19\%) + CFCl_2CF_2CCl_3\ (20\%) + CF_2ClCCl_2CFCl_2\ (3\%) \quad \text{CF/CCl 86:14}$$

$$CCl_3F \xrightarrow[19\text{–}21°C]{CClF{=}CF_2,\ AlCl_3} CF_2ClCFClCCl_3\ (37.4\%) + CFCl_2CFClCFCl_2\ (16.5\%) \quad \text{CF/CCl 70:30}$$

Similar addition takes place when fluoro-olefins are treated with acyl chlorides in the presence of aluminum chloride **[359]**.

(199) $$CH_3COCl + CHF{=}CF_2 \xrightarrow[-5\ \text{to}\ -10°C]{AlCl_3,\ CHCl_3} CH_3COCHFCF_2Cl \quad 34\%$$ **[359]**

*Acid-Catalyzed Substitutions*

The domain of acid-catalyzed syntheses is aromatic series, where substitution occurs exclusively. Fluorinated compounds can act as substrate (passive components), or as reagents (active components), or both.

Fluorobenzene gives *Friedel-Crafts reactions* with various reagents including chloroparaffins **[360]**, aldehydes **[361,362]**, and acyl chlorides **[363]**. Substitution takes place in *p*-position to fluorine. Even strongly deactivated aromatic systems such as pentafluorobenzene undergo the reaction **[364]**.

(200)

$$C_6H_5F \xrightarrow[CS_2,\ 15°C]{CCl_4,\ AlCl_3} (p\text{-}FC_6H_4)_2CCl_2 \quad 59\% \quad [360]$$

$$C_6H_5F \xrightarrow[HCl,\ ZnCl_2]{(CH_2O)n} p\text{-}FC_6H_4CH_2Cl \quad [361]\ 60.3\%$$

$$C_6H_5F \xrightarrow[H_2SO_4]{CCl_3CHO} (p\text{-}FC_6H_4)_2CHCCl_3 \quad 33\% \quad [362]$$

$$C_6H_5F \xrightarrow[CS_2]{CH_3COCl,\ AlCl_3} p\text{-}FC_6H_4COCH_3 \quad 76\% \quad [363]$$

(201) $$C_6HF_5 \xrightarrow[150°C,\ 4.5\ \text{hr, autoclave}]{CHCl_3,\ AlCl_3} CH(C_6F_5)_3 \quad 92\%$$ **[364]**

When fluorohaloparaffins are used as reagents in Friedel-Crafts alkylation of aromatics, the carbon–fluorine bond is cleaved preferentially to the carbon–halogen bond. Consequently, fluorine-free products result **[365]**.

(202)
$$C_6H_5CH(CH_3)CH_2Cl\ 92\% \xleftarrow[BF_3,\ -10\ to\ 10°C]{CH_3CHFCH_2Cl} C_6H_6 \xrightarrow[BF_3,\ 0\text{–}20°C]{CH_3CHFCH_2Br} C_6H_5CH(CH_3)CH_2Br\ 89\%$$
[365]

The acylation of aromatics with fluorinated acyl chlorides is a good synthetic route to aromatic-aliphatic ketones **[366–369]**.

(203)
$$C_6H_5COCH_2F\ 81\%\ [366] \xleftarrow[AlCl_3,\ CH_2Cl_2,\ 0°C]{FCH_2COCl} C_6H_6 \xrightarrow[AlCl_3,\ 10\text{–}20°C]{CF_3COCl} C_6H_5COCF_3\ 64\%\ [368]$$
$$C_6H_6 \xrightarrow{CHF_2COCl,\ AlCl_3} C_6H_5COCHF_2\ 69\%\ [367]$$

(204)
$$C_6F_5CH_2COCl \xrightarrow[-10°C]{C_6H_5CH_3,\ FeCl_3} p\text{-}CH_3C_6H_4COCH_2C_6F_5\ 61\%$$
[369]

A peculiar acid-catalyzed reaction occurs between aromatic hydrocarbons and perchloryl fluoride in the presence of aluminum chloride. A new type of aromatic compounds is formed. The perchloryl group attached to the benzene ring is a strongly *meta*-orienting substituent **[370]**.

(205)
$$C_6H_6 \xrightarrow[40°C]{FClO_3,\ AlCl_3} C_6H_5ClO_3\ 40\%$$
[370]

## NUCLEOPHILIC SUBSTITUTIONS

In this section on nucleophilic substitutions, the following reactions will be discussed: esterification, hydrolysis, alkylation, arylation, acylation (Table 44), syntheses with organometallics, and base-catalyzed condensations.

### Esterification and Acetalization

Both esterification and acetalization are acid-catalyzed nucleophilic reactions of great preparative importance.

The *esterification* of polyfluorinated alcohols, which are very acidic, gives poor yields unless trifluoroacetic anhydride is used as a catalyst. It evidently combines with acids to form mixed anhydrides, which react via a carbonium mechanism **[371]**.

**Table 44. Classification of Nucleophilic Displacement Reactions**

$$Z\text{—}\overset{\ominus}{Y} + R\text{—}X^{a} \rightarrow Z\text{—}Y\text{—}R + \overset{\ominus}{X}$$

| Y | Z = H | R = Alkyl | R = Aryl | R = Acyl |
|---|---|---|---|---|
| $-\overset{\ominus}{O}$ | Hydrolysis | O-Alkylation | O-Arylation | O-Acylation |
| $-\overset{\ominus}{S}$ | — | S-Alkylation | S-Arylation | S-Acylation |
| $-\overset{\ominus}{N}H$, $-\overset{\ominus}{N}-$ | Ammonolysis | N-Alkylation | N-Arylation | N-Acylation |
| $-\overset{\ominus}{C}-$ | — | C-Alkylation | C-Arylation | C-Acylation |

[a] X=F, Cl, Br, or I.

(206) $C_3F_7CH_2OH + CH_2{=}CHCO_2H \xrightarrow[30°C]{(CF_3CO)_2O} C_3F_7CH_2OCOCH{=}CH_2$ 91.5% **[371]**

The esterification of trifluoroacetic acid, which is strongly acidic, would take place even without a catalyst. Nevertheless, sulfuric acid is usually added to the reaction mixture **[372,373]**.

(207) $CF_3CO_2H \xrightarrow{C_2H_5OH,\ H_2SO_4,\ distn}$ 82% $CF_3CO_2C_2H_5$ **[372]**

$CF_3CO_2K \xrightarrow{C_2H_5OH,\ H_2SO_4,\ distn}$ 73% $CF_3CO_2C_2H_5$ **[373]**

Other methods of preparing esters from fluorinated acids are the reactions of acids with diazoalkanes, and transesterification **[374]**.

(208) $CF_3CO_2H + P(OC_6H_5)_3 \xrightarrow[distn\ 250–300°C]{2.5\ hr\ reflux} CF_3CO_2C_6H_5$ 56% **[374]**

Esters of fluorinated acids can also be prepared from fluorinated amides **[375,376]** or nitriles **[377]**, which are sometimes more readily accessible than the free acids.

(209) $CH_2FCONH_2 + C_2H_5OH \xrightarrow{HCl(g),\ 20°C,\ 36\ hr}$ 66% $CH_2FCO_2C_2H_5$ **[375]**

$CH_2FCONH_2 + C_2H_5OH \xrightarrow{H_2SO_4,\ reflux\ 4.5\ hr}$ 67.5% $CH_2FCO_2C_2H_5$ **[376]**

(210) $F(CH_2)_4CN \xrightarrow[reflux\ 1\ hr]{CH_3OH,\ HCl(g)} F(CH_2)_4CO_2CH_3$ 83.5% **[377]**

The *acetalization* of fluorinated aldehydes and ketones is accomplished in a common way using sulfuric acid as a catalyst **[378]**.

(211) $$CH_3COCF_3 + C_3H_5(OH)_3 \xrightarrow[\text{4 days}]{H_2SO_4} (CF_3)(CH_3)C\langle OCH_2—OCHCH_2OH \rangle \quad 65\text{–}70\%$$ [378]

## Hydrolysis

The hydrolysis of fluorinated compounds may or may not result in displacement of fluorine. The stability of fluorine atoms toward hydrolysis depends mainly on the position in the organic molecule and on reaction conditions.

### *Hydrolysis of Nonfluorinated Parts of Fluorinated Molecules*

Fluorinated esters, amides, and nitriles give fluorinated acids. If fluorine atoms are located in such positions that they can be intramolecularly displaced by the newly formed carboxylic groups, very mild conditions of hydrolysis (dilute alkalies, low temperature, short time of reaction) must be used if elimination of fluorine is to be prevented.

In fluorohaloparaffins, halogen atoms are displaced preferentially to fluorine in acid hydrolysis. Such reactions represent a convenient method for the preparation of fluoro- or fluorohalocarboxylic acids **[379]**.

(212) $$CF_2BrCBr_3 \xrightarrow[\text{100°C, 2 hr, reflux overnight}]{\text{65\% oleum, HgSO}_4\text{, Hg}_2\text{SO}_4} CF_2BrCO_2H \quad 61\%$$

$$CF_2BrCBr_3 \xrightarrow[\text{100°C, 2 hr, reflux; 2: C}_2\text{H}_5\text{OH}]{\text{1: 30\% oleum, HgSO}_4\text{, Hg}_2\text{SO}_4} CF_2BrCO_2C_2H_5 \quad 60.5\%$$ [379]

### *Hydrolytic Displacement of Single Fluorine Atoms*

Single fluorine atoms in saturated compounds are fairly stable toward hydrolysis unless their position in the molecule allows an intramolecular attack of a hydroxylic or carboxylic group (p. 87).

Single fluorine atoms bound to double-bond carbons are readily removed by secondary reactions, the primary reactions being addition of a nucleophile across the double bond **[380]**.

(213) $$CF_2\langle CF_2—CF=CF—CF_2 \rangle \xrightarrow[\text{50°C, 24 hr}]{OH^\ominus} \left[ CF_2\langle CF_2—\overset{\ominus}{C}F—CFOH—CF_2 \rangle \right] \xrightarrow[OH^\ominus]{-F^\ominus} CF_2\langle CF=CF—C(OH)_2—CF_2 \rangle$$

$$\xrightarrow[OH^\ominus]{-F^\ominus} CF_2\langle CF_2—CO—C(OH)=CF \rangle \rightleftarrows CF_2\langle CO—CF=C(OH)—CF_2 \rangle \quad 62.5\%$$ [380]

A fluorine atom bound to an aromatic nucleus is resistant to hydrolysis unless its nucleophilic displacement is facilitated by electron-withdrawing groups, especially nitro groups, in *ortho* and/or *para* positions **[381]**, or an overall electron-poor nature of the ring, as in perfluoroaromatics **[382]**.

(214) 2-fluoro-3-nitro-5-nitro... $O_2N$–$C_6H_2$($CO_2H$)($NO_2$)–F $\xrightarrow[20°C,\ 10\ min]{NaOH}$ $O_2N$–$C_6H_2$($CO_2H$)($NO_2$)–OH 90% **[381]**

(215) $C_6F_6 \xrightarrow[reflux\ 1\ hr]{KOH,\ (CH_3)_3COH} C_6F_5OH$ 71% **[282]**

Nucleophilic displacement of aromatic fluorine may take place preferentially to chlorine, especially if fluorine atoms are activated by proper positions in the nucleus **[383]**.

(216) 3,5-dichloro-2,4,6-trifluoropyridine $\xrightarrow[reflux\ 1.5\ hr]{5\%\ KOH,\ (CH_3)_3COH}$ or $\xrightarrow[80°C,\ 20\ hr]{5\%\ KOH,\ H_2O,\ sealed}$ 3,5-dichloro-2,6-difluoro-4-hydroxypyridine (25.5%; 76.5%) + 3,5-dichloro-4,6-difluoro-2-hydroxypyridine (59.5%; 8.5%) **[383]**

Fluorine bound to silicon and phosphorus is readily replaced by hydroxyl even under very mild conditions **[384]**.

(217) $[(CH_3)_2O]_2P(O)F \xrightarrow[15°C,\ 72\ hr]{H_2O} [(CH_3)_2O]_2P(O)OH$ **[384]**

*Hydrolysis of Difluoromethylene Group*

The geminal difluoro grouping is readily hydrolyzed, especially in an acidic medium, provided the difluoromethylene group is adjacent to a double bond **[385–387]**, or nitrogen **[388]**, or oxygen atoms **[389,390]**. Acid hydrolysis of the difluoromethylene group in 4,4-dichloro-3,3-difluoro-1-phenylcyclobutene takes place preferentially to the dichloromethylene group **[386]**.

(218) $C_6H_5C{<}(CH_2)(CF{=})CF_2$ $\xrightarrow[C_2H_5OH]{KOH}$ $C_6H_5C{<}(CH_2)(=C(OC_2H_5))CF_2$ 68% $\xrightarrow[100°C]{92\%\ H_2SO_4}$ $C_6H_5C{<}(CH_2)(=C(OC_2H_5))CO$ 74% **[385]**

(219) $C_6H_5C{<}(CH{=})(CCl_2)CF_2$ $\xrightarrow[100°C,\ 11\ min]{H_2SO_4}$ $C_6H_5C{<}(CH{=})(CCl_2)CO$ 82% **[386]**

(220) $(CF_2)_n$ ring with $CF_2$–CCl=CCl–$CF_2$ $\xrightarrow[85°C]{SO_3,\ SbF_5}$ $(CF_2)_n$ ring with $CF_2$–CCl=CCl–CO + $(CF_2)_n$ ring with CO–CCl=CCl–CO [387]

for $n=1$: 2.5 hr 48% 35%
for $n=2$: 20 hr 78% 19%

(221) $CHF_2CF_2N(C_2H_5)_2 \xrightarrow[ice]{H_2O} CHF_2CON(C_2H_5)_2$ 95% [388]

(222)
$CHClFCF_2OC_2H_5 \xrightarrow[0\text{–}10°C]{96\%\ H_2SO_4}$ 83% $CHClFCO_2C_2H_5$ [389]

$CHClFCO_2C_2H_5$ ↓ NaOH; ↑ $C_2H_5OH$

$CHClFCF_2OC_6H_5 \xrightarrow[100°C]{96\%\ H_2SO_4}$ 45% $CHClFCO_2H$ [390]

*Hydrolysis of Trifluoromethyl Group*

Trifluoromethyl groups are resistant to alkaline hydrolysis except in cases in which elimination of hydrogen fluoride creates a double bond susceptible to an attack by a nucleophile. A combined addition–elimination mechanism results finally in the total displacement of fluorine and the formation of a carboxylic group **[391]**.

(223) [391]

$CF_3CH(CH_3)CO_2H \xrightarrow[100°C]{2N\ NaOH} [CF_2{=}C(CH_3)CO_2H \longrightarrow HOCF_2CH(CH_3)CO_2H] \longrightarrow HOCOCH(CH_3)CO_2H$ 62%

Strong acids hydrolyze the trifluoromethyl group to a carboxylic group under vigorous conditions **[392–394]**.

(224)

$C_6H_5CF_3 \xrightarrow[160°C]{HBr(d\ 1.79),\ SiO_2} C_6H_5CO_2H$ quant. [392]

$C_6H_5CF_3 \xrightarrow[heat]{100\%\ H_2SO_4}$ 94% $C_6H_5CO_2H$ [393]

$C_6H_5CF_3 \xrightarrow[100°C]{HF,\ H_2O} C_6H_5CO_2H$ quant. [394]

*Hydrolysis of Perfluoro Compounds*

Saturated perfluoro compounds are extremely resistant to any kind of hydrolysis. On the other hand, perfluoro-olefins are susceptible to nucleophilic additions, which lead to intermediates readily hydrolyzed by alkalies or acids according to the following scheme **[395]**:

(225) [395]

$CF_3CF{=}CF_2 \xrightarrow[MeOH]{KOH} CF_3CHFCF_2OCH_3 \xrightarrow[10\text{–}20°C]{H_2SO_4} CF_3CHFCO_2CH_3 \xrightarrow{H_3PO_4} CF_3CHFCO_2H$

In perfluoroaromatics, fluorine in the nucleus is easily displaced by hydroxyl in alkaline hydrolysis. Acid hydrolysis hydrolyzes the trifluoromethyl group in perfluorotoluene to a carboxylic group **[396]**.

(226) $$C_6F_5CF_3 \xrightarrow[\text{reflux 1 week}]{H_2SO_4} C_6F_5CO_2H \quad 25\%$$ **[396]**

*Fluoroform Reaction*

Trifluoromethyl and generally perfluoroalkyl ketones are cleaved by strong alkalies to acids and fluoroform or 1H-perfluoroalkane, respectively **[397,398]**.

(227) $$C_6H_5COCF_3 \xrightarrow{10\%\ KOH} C_6H_5CO_2H + CHF_3$$ **[397]**

(228) $$C_3F_7CO_2C_2H_5 \xrightarrow[\text{reflux}]{\text{EtONa, EtOH}} \underset{71\%}{C_3HF_7} + \underset{30\%}{CO(OC_2H_5)_2}$$ **[398]**

## Alkylations

Alkylations can be considered as nucleophilic displacements of (mainly) halogens in aliphatic chains by oxygen, sulfur, nitrogen, or carbon nucleophiles. Fluorine is usually more resistant to such displacement reactions than other halogens. A comparison is made in Table 45 **[399]**.

*Alkylations at Oxygen*

Because of the lower reactivity of fluorine, other halogens are replaced preferentially when fluorohalo compounds react with oxygen nucleophiles **[400]**. However, even fluorine can be replaced **[401]**. Such a replacement is especially easy in intramolecular alkylations **[275]**.

(229) 2,4-dichlorophenol sodium salt (ONa, Cl, Cl on benzene ring) $+ ClCHFCO_2C_2H_5 \xrightarrow[\text{reflux 2 hr}]{\text{EtOH}}$ 2,4-dichlorophenoxy compound ($OCHFCO_2C_2H_5$, Cl, Cl on benzene ring) 42.7% **[400]**

**Table 45. Relative Reactivities of Isoamyl Halides and Halobenzenes toward Sodium Methoxide and Piperidine at 18°C [399]**

| | | Relative reactivity of | | | |
|---|---|---|---|---|---|
| Substrate | Reagent | F | Cl | Br | I |
| $(CH_3)_2CHCH_2CH_2X$ | $CH_3ONa$ | 1 | 71 | 3,550 | 4,500 |
| | $C_5H_{11}N$ | 1 | 68.5 | 17,800 | 50,500 |
| $C_6H_5X$ | $CH_3ONa$ | 1 | 1.8 | 4.4 | 35.6 |
| | $C_5H_{11}N$ | 1 | 1.9 | 74.5 | 132 |

(230) $C_6H_5OCH_2CH_2OH$ 55% $\xleftarrow[\text{30\% NaOH, reflux 0.5 hr}]{C_6H_5ONa}$ $FCH_2CH_2OH$ $\xrightarrow[\text{30\% NaOH, reflux 0.5 hr}]{C_6H_5SNa}$ 63.5% $C_6H_5SCH_2CH_2OH$ [401]

In unsaturated fluorinated compounds, direct replacement of halogens is rare. More often, addition–elimination takes place, as proved by the isolation of addition products of fluorohalo-olefins and alcohols or phenol at milder conditions. In agreement with the mechanism, fluorine atoms bound to double bonds are replaced preferentially to other halogens at the same double bond **[402]**.

(231) $CHCl_2CF_2OC_6H_5$ 60% $\xleftarrow[\text{KOH, 10°C}]{C_6H_5OH}$ $CCl_2{=}CF_2$ $\xrightarrow[\text{KOH, 70°C}]{C_6H_5OH}$ 62% $CCl_2{=}CFOC_6H_5$ [402]

With higher fluoro- and fluorohalo-olefins, the situation is rather complicated since the reaction can take place either by the $S_N2$ or by the $S_N2'$ mechanism. The following examples are illustrative **[403,404]**:

(232) $CF_2$—CH=CH—$CF_2$ (cyclic) $\xrightarrow{\text{1 KOH, EtOH}}$ $CF_2$—CH=CH—CF($OC_2H_5$) (cyclic) 74.5% $\xrightarrow{\text{1 KOH, EtOH}}$ $CF_2$—CH=CH—C($OC_2H_5$)($OC_2H_5$) (cyclic) 82.5% [403]

(233) $CF_2$—CH=CCl—$CF_2$ (cyclic) $\xrightarrow[\text{0°C}]{\text{KOH, EtOH}}$ 77%; $CF_2$—CH=CCl—CFCl (cyclic) $\xrightarrow[\text{0°C}]{\text{KOH, EtOH}}$ 92%; both give $CF_2$—$CHOC_2H_5$—CCl=CF (cyclic) $\xrightarrow{\text{KOH, EtOH, 0°C}}$ $CF_2$—$CHOC_2H_5$—CCl=C$OC_2H_5$ (cyclic) 82% [403]

(234) cyclo-$CF_2CF_2CF_2$CCl=CCl $\xrightarrow{\text{KOH, EtOH}}$ cyclo-$CF_2CF_2CF_2$CCl=$COC_2H_5$ 47% $\xrightarrow{\text{KOH, EtOH}}$ cyclo-$CF_2CF_2$C($OC_2H_5$)=CCl—C($OC_2H_5$)$_2$ 49% [404]

From the enormous experimental data, especially in the field of cyclic polyfluorohalo-olefins, the following rules can be derived to help estimate the probable results of a reaction **[405]**:

1. The ease of the replacement of halogens decreases from fluorine to bromine. Since the mesomeric shift of electrons by fluorine is stronger than that caused by chlorine or bromine, the nucleophile joins the less electronegative end of the double bond, i.e., the carbon atom carrying fluorine. The double bond is regenerated by the elimination of fluorine.

(235) Mesomeric effect F > Cl > Br **[405]**

$$(F)C{=}C(Cl) \longleftrightarrow (F)\overset{\oplus}{C}{-}\overset{\ominus}{\ddot{C}}(Cl) \xrightarrow{X^{\ominus}} (F)(X)C{-}\overset{\ominus}{\ddot{C}}(Cl) \xrightarrow{-F^{\ominus}} (X)C{=}C(Cl)$$

2. When both halogens at the double bond are the same, the inductive effect of the neighboring groups determines the direction of the nucleophilic attack. The nucleophile joins the carbon atom more distant from the more electronegative grouping. The order of decreasing inductive effect is

Inductive effect

$$Cl_2 > Cl,\ OEt > ClF > (OEt)_2 > CF_2 > CH_2$$

(236) **[405]**

$$\begin{matrix} CCl_2{-}CF_2 \\ | \quad\quad | \\ CCl{=}CCl \end{matrix} \longleftrightarrow \begin{matrix} CCl_2{-}CF_2 \\ | \quad\quad | \\ \ddot{C}Cl{-}CCl \end{matrix} \xrightarrow[-Cl^{\ominus}]{X^{\ominus}} \begin{matrix} CCl_2{-}CF_2 \\ | \quad\quad | \\ CCl{=}CX \end{matrix} \gg \begin{matrix} CCl_2{-}CF_2 \\ | \quad\quad | \\ CX{=}CCl \end{matrix}$$

3. Finally, the leaving ability of the corresponding leaving groups also determines the outcome of the reaction. The leaving ability decreases in the sequence.

(237) Leaving ability **[405]**

$$C\langle^{OEt}_{Br} > C\langle^{Cl}_{Cl} > C\langle^{OEt}_{Cl} > C\langle^{F}_{Cl} > C\langle^{F}_{F} \gg C\langle^{OEt}_{OEt},\ C\langle^{H}_{H}$$

Never leave

Since a strong mesomeric effect usually parallels poor leaving ability, it is difficult to distinguish which factor is more responsible.

*Alkylations at Sulfur*

In saturated chains, other halogens react preferentially to fluorine with sulfur nucleophiles in direct displacement reactions **[406,407]**.

(238) **[406]**

$$F(CH_2)_6Br \xrightarrow[\text{reflux}]{\text{KSCN, EtOH}} F(CH_2)_6SCN \quad 95\%$$

(239) $$CHClF_2 \xrightarrow[\text{7 hr}]{C_6H_5SH,\ MeONa} CHF_2SC_6H_5 \quad 63\%$$ **[407]**

In unsaturated fluorohalo compounds with fluorine and halogen atoms at the double bonds, the direct replacement is simulated by addition–elimination reactions. Under special conditions, addition products can be isolated **[408]**.

(240) **[408]**

$$\text{CF=CF-CF}_2\text{-CF}_2 \text{ (cyclobutene)} \xrightarrow[\text{44°C, 7 days, 65°C, 0.5 hr}]{C_4H_9SH,\ \text{Triton B}} \text{C}_4\text{H}_9\text{SCH-CFSC}_4\text{H}_9\text{ (ring with CF}_2\text{-CF}_2\text{)} \quad 30\%$$

$$\text{CF=CF-CF}_2\text{-CF}_2 \xrightarrow[<0°C]{C_4H_9SH,\ Et_3N} \text{C}_4\text{H}_9\text{SC=CF (ring with CF}_2\text{-CF}_2\text{)}\ 23\% + \text{C}_4\text{H}_9\text{SC=CSC}_4\text{H}_9\text{ (ring with CF}_2\text{-CF}_2\text{)}\ 23\%$$

*Alkylations at Nitrogen and Phosphorus*

In the reaction of saturated fluorohalo compounds, both direct replacement or elimination–addition can take place **[409]**. In unsaturated fluorohalo-olefins, $S_N2$ and $S_N2'$ mechanisms account best for the results of the reactions with nitrogen nucleophiles **[410,411]**.

(241) **[409]**

$$CF_3CH_2CHBrCO_2C_2H_5 \xrightarrow[\text{reflux 24 hr}]{NaN_3,\ EtOH} CF_3CH_2CHN_3CO_2C_2H_5\ 63\% \xrightarrow{H_2} CF_3CH_2CH(NH_2)CO_2C_2H_5\ 96\%$$

$$CF_3CH_2CHBrCO_2C_2H_5 \xrightarrow{NH_3} [CF_3CH{=}CHCO_2C_2H_5] \longrightarrow CF_3CH(NH_2)CH_2CONH_2\ 42\%$$

(242) $$(CF_3)_2C{=}CF_2 \xrightarrow[0°C]{NH(C_2H_5)_2,\ Et_2O} (CF_3)_2C{=}CFN(C_2H_5)_2 \quad 65\%$$ **[410]**

(243) $$\text{CF}_2\text{-CCl=CCl-CF}_2 \text{ (ring)} \xrightarrow[\text{0°C, 2 hr}]{CH_3NH_2,\ Et_2O} \left[\text{CF}_2\text{-CClNHCH}_3\text{, CF=CCl (ring)}\right] \longrightarrow \text{CF}_2\text{-C=NCH}_3\text{, CF=CCl (ring)} \quad 69\%$$ **[411]**

The alkylation of triethyl phosphite with 1-chloroperfluorocyclopentene gives, via Arbusov rearrangement, diethyl 2-chloroperfluorocyclopentenephosphonate. Again, fluorine bound to the double bond is replaced preferentially to chlorine **[412]**.

(244) $$\text{CCl=CF-CF}_2\text{-CF}_2\text{-CF}_2 \text{ (ring)} \xrightarrow[25\text{–}100°C]{P(OC_2H_5)_3} \text{CCl=CP(O)(C}_2\text{H}_5)_2\text{-CF}_2\text{-CF}_2\text{-CF}_2 \text{ (ring)} \quad 33\%$$ **[412]**

*Alkylations at Carbon*

Alkylations at carbon occur in the reaction of fluorohaloalkanes with cyanide **[413]** and with acetylide **[414]**, and in the reaction of alkyl or benzyl halides with ethyl fluoroacetate **[415]**.

(245) $$F(CH_2)_6Br \xrightarrow[\text{reflux 7 hr}]{\text{NaCN, 80\% EtOH}} F(CH_2)_6CN \quad 90\%$$ **[413]**

(246) $$F(CH_2)_4Br \xrightarrow{\text{NaC}\equiv\text{CH, NH}_3(l)} F(CH_2)_4C\equiv CH \quad 89\%$$ **[414]**

(247) **[415]**
$$C_6H_5CH_2Br + CH_2FCO_2C_2H_5 \xrightarrow{\text{EtONa, EtOH}} C_6H_5CH_2CHFCO_2C_2H_5 \quad 27\%$$
$$C_6H_5CH_2Br + CH_2FCO_2C_2H_5 \xrightarrow{\text{NaH, Et}_2\text{O}} C_6H_5CH_2CHFCO_2C_2H_5 \quad 20\%$$

A peculiar alkylation at carbont akes place when 1,4-diiodoperfluorobutane is heated with benzene in the presence of sodium acetate. The resulting octafluorotetralin can be converted by defluorination to tetrafluoronaphthalene **[416]**.

(248) **[416]**
$$C_6H_6 + ICF_2CF_2CF_2CF_2I \xrightarrow[200^\circ C]{CH_3CO_2Na} C_6H_4(CF_2)_4 \xrightarrow[500^\circ C]{Fe} C_{10}H_4F_4$$

## Arylations

Nucleophilic displacements of fluorine and other halogens in aromatic nuclei are generally more difficult than those in the aliphatic series (Table 45). Contrary to the case of saturated fluorohalo derivatives, and similar to that of unsaturated fluorohalo compounds, aromatic fluorine is replaced preferentially to other halogens in nucleophilic reactions. Evidently the strong mesomeric effect of fluorine causes greater electron dilution at the carbon at which it is bound than does the same effect when due to the other halogens.

*Arylations at Oxygen*

Displacement of fluorine in the benzene ring is especially easy when electron-withdrawing groups such as carboxylic groups or nitro groups are present in *ortho* and/or *para* positions **[417]**.

(249) [417]

X, Y, $NO_2$ $\xrightarrow{C_2H_5ONa}$ $OC_2H_5$, Y, $NO_2$

| Relative reactivity for | | | | |
|---|---|---|---|---|
| X = | F | Cl | Br | I |
| Y = H | 228 | 1 | 0.87 | 0.74 |
| Y = $NO_2$ | 581 | 1 | | |

Polyfluorohalobenzene derivatives and perfluorobenzene and other perfluoro aromatics readily undergo nucleophilic reactions in which fluorine is replaced in preference to other halogens. The ease of the displacement is accounted for by strong electron dilution at the nucleus caused by the inductive effect of many fluorine atoms. Perfluorobenzene and sodium methoxide give pentafluoroanisole. In bromo- and iodopentafluorobenzene, fluorine atoms *para* to bromine or iodine are replaced preferentially **[418–421]**.

(250) [418]

Bromopentafluorobenzene $\xrightarrow[\text{reflux 15 hr}]{CH_3ONa,\ C_5H_5N}$ 4-bromo-2,3,5,6-tetrafluoroanisole ($OCH_3$), 51%

Iodopentafluorobenzene $\xrightarrow[\text{reflux 3 hr}]{CH_3ONa,\ C_5H_5N}$ 4-iodo-2,3,5,6-tetrafluoroanisole ($OCH_3$), 53%

(251)

Perfluorobenzene (F):

$\xrightarrow[\text{reflux 6 hr}]{CH_3ONa,\ CH_3OH}$ 72% [419]

$\xrightarrow[\text{reflux 1 hr}]{C_2H_5OK,\ C_2H_5OH}$ 56% [421]

$\xrightarrow[80°C,\ 0.5\ hr]{CH_3ONa,\ CH_3OH,\ C_5H_5N}$ $OCH_3$ (F), 70% [420]

$\xrightarrow[120°C,\ 0.5\ hr]{C_2H_5OH,\ KOH}$ $OC_2H_5$ (F), 48% [421]

In fluorinated aromatic nitrogen-containing heterocycles, fluorine is replaced preferentially in *ortho* and/or *para* positions to the nitrogen since these positions are activated for nucleophilic attacks **[422]**.

(252) [422]

Tetrafluoropyrimidine:

$\xrightarrow[20°C,\ 0.5\ hr]{CH_3OH,\ Na_2CO_3}$ $OCH_3$, F, N, F, N, F — 52%

$\xrightarrow[\text{reflux}]{CH_3ONa,\ CH_3OH}$ $OCH_3$, F, N, $CH_3O$, N, $OCH_3$ — 51%

$\xrightarrow[THF,\ 0\ °C]{CH_3ONa,\ CH_3OH}$ $OCH_3$, F, N, $CH_3O$, N, F — 34%

*Arylations at Sulfur*

Arylations at sulfur occur in reactions of fluorinated aromatics with sulfur nucleophiles **[423,424]**. Where there is a choice between oxygen and sulfur as nucleophile, the reaction occurs at sulfur exclusively **[425]**. In pentafluorobenzene, the sulfhydryl group preferentially replaces fluorine *para* to the hydrogen atom **[423]**.

(253) $C_6F_6$ $\xrightarrow[(CH_2OH)_2,\ \text{reflux 5 min}]{H_2S,\ NaOH,\ C_5H_5N}$ $C_6F_5SH$ (66%) $\xrightarrow[\text{BuOH, reflux 6 hr}]{\text{Raney Ni}}$ $C_6F_5H$ (60%) $\xrightarrow[\text{reflux 5 min}]{NaSH,\ C_5H_5N}$ $p\text{-}HC_6F_4SH$ 48% **[423]**

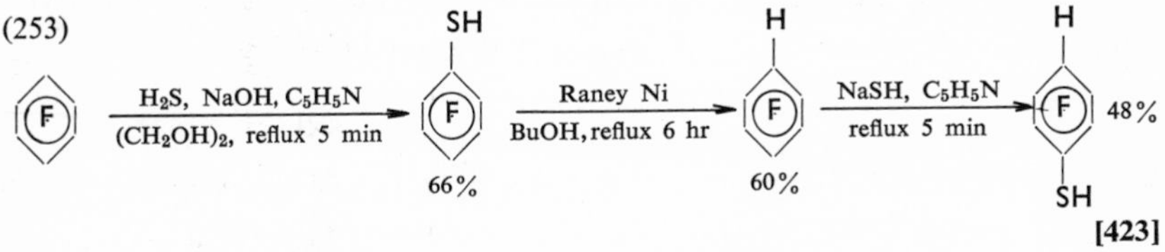

(254) $C_6F_6$ $\xrightarrow[\text{reflux, 10 min}]{C_6H_5SK,\ C_5H_5N}$ $p\text{-}C_6H_5SC_6F_4SC_6H_5$ 55% **[424]**

$C_6F_6$ $\xrightarrow[\text{dioxane, 14 hr}]{HOCH_2CH_2SH,\ K_2CO_3}$ $p\text{-}HOCH_2CH_2SC_6F_4SCH_2CH_2OH$ 43% **[425]**

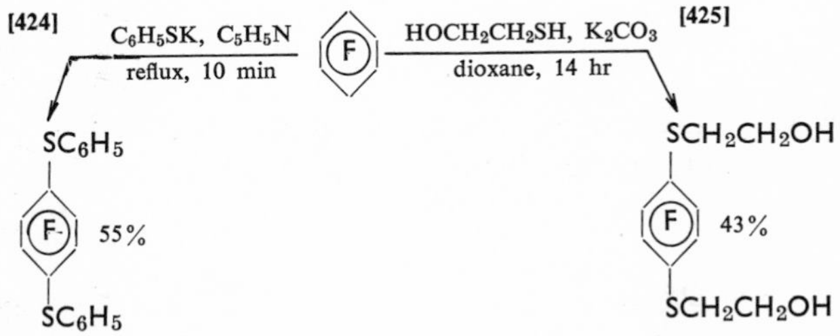

*Arylations at Nitrogen*

Arylations at nitrogen are reactions of fluorinated aromatics with ammonia, amines, and hydrazine. Perfluorobenzene reacts very readily, and, strangely enough, *m*-derivatives prevail in products when an excess of the nucleophile is used **[426,427]**.

(255) $C_6F_6$ $\xrightarrow[150°C,\ 3\ \text{hr}]{30\%\ NH_4OH}$ $C_6F_5NH_2$ 85% $\rightarrow$ $m\text{-}H_2NC_6F_4NH_2$ 76.5–81% + $p\text{-}H_2NC_6F_4NH_2$ 9% **[426]**

$C_6F_6$ $\xrightarrow[\text{reflux, 45 hr}]{N_2H_4,\ THF}$ $C_6F_5NHNH_2$ 32% $\rightarrow$ $p\text{-}H_2NHNC_6F_4NHNH_2$ < + $m\text{-}H_2NHNC_6F_4NHNH_2$ (35%) **[427]**

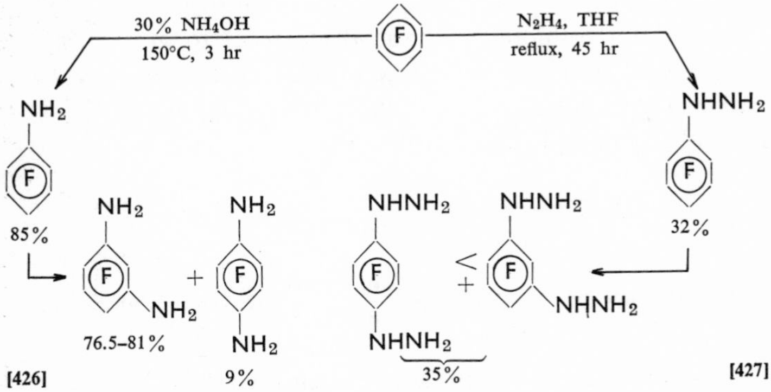

In chlorofluoropyridine, fluorine in $\gamma$-position is replaced most readily, even in preference to chlorine in $\beta$-position **[428]**.

(256) [structure: 3,5-dichloro-2,4,6-trifluoropyridine] $\xrightarrow[\text{dioxane } 10\text{–}15°C]{N_2H_4,\ H_2O}$ [structure: 4-hydrazino-3,5-dichloro-2,6... pyridine with $NHNH_2$, Cl, Cl, F] 90% [428]

*Arylations at Carbon*

The displacement of aromatic fluorine in perfluorobenzene by carbanions is very rare and, so far, of little practical value **[429]**.

(257) $$C_6F_6 + CH_2(CN)CO_2C_2H_5 \xrightarrow[\text{reflux, 5 hr}]{\text{NaH, HCONMe}_2} C_6F_5CH(CN)CO_2C_2H_5$$ [429]

## Acylations

Acylations can occur at oxygen, sulfur, nitrogen, and carbon. The best acylating agents are acyl halides and acid anhydrides. Esters are used less frequently.

*Acylations at Oxygen*

Acylations at oxygen take place when alcohols or phenols are treated with acyl halides or anhydrides **[430,431]**.

(258) [430]
$$C_5F_{11}CH_2OH + ClCOCH{=}CH_2 \xrightarrow[0\text{–}67°C,\ 5\ min]{p\text{-}C_6H_4(OH)_2} C_5F_{11}CH_2OCOCH{=}CH_2 \quad 82.5\%$$

(259) $$C_6H_5OH + ClCOC_3F_7 \xrightarrow[6\ hr]{190°C} C_6H_5OCOC_3F_7 \quad 52.5\%$$ [431]

Phenol and trifluoroacetic anhydride react to give phenyl trifluoroacetate, a very good acylating agent **[432]**. The compound can also be prepared by decomposition of a mixed anhydride obtained from trifluoroacetic acid and phenyl chloroformate **[433]**.

(260)
$$C_6H_5OH + (CF_3CO)_2O \xrightarrow{120°C} 95\%\ C_6H_5OCOCF_3$$ [432]

$$C_6H_5OCOCl + CF_3CO_2H \xrightarrow[\text{reflux, 1 hr}]{Et_3N,\ THF} [C_6H_5OCOOCOCF_3] \xrightarrow{75\text{–}80\%} C_6H_5OCOCF_3$$ [433]

Trifluoroacetylation at oxygen is often used in carbohydrate chemistry for temporary blocking of hydrolytic groups. Regeneration of free hydroxyl groups is achieved at very mild conditions, usually by transesterification with methanol at room temperature **[434]**.

*Acylations at Sulfur*

The reaction of trifluoroacetic anhydride with hydrogen sulfide gives trifluorothiolacetic acid **[435]**. The reaction of trifluoroacetyl chloride with silver thiocyanate gives trifluoroacetyl thiocyanate **[436]**.

(261) $$CF_3COCl \xrightarrow[60\text{–}80°C,\ 2\ days]{AgSCN} CF_3COSCN\ \ 60\%$$ **[436]**

*Acylations at Nitrogen*

Acylations at nitrogen take place in reactions of acyl halides, acid anhydrides, or esters with ammonia, amines, hydrazine **[438]**, or azide **[437]**.

(262) $$C_3F_7COCl \xrightarrow[55\text{–}65°C,\ 24\ hr]{NaN_3,\ C_6H_6} C_3F_7CON_3\ \ {>}75\%\quad [437]$$

$$C_3F_7COCl \xrightarrow[reflux]{N_2H_4 \cdot H_2O,\ C_6H_6} C_3F_7CONHNHCOC_3F_7\ \ 59\%\quad [438]$$

Trifluoroacetylation is of special importance in the field of amino acids. Different amino groups in the same molecule can be acylated, depending on the reagent used (trifluoroacetic anhydride, ethyl trifluorothiolacetate, phenyl trifluoroacetate). An example is selective acylation of lysine **[439,440]**.

(263) $$H_2NCH_2CH_2CH_2CH_2CH(NH_2)CO_2H \xrightarrow[CF_3CO_2H,\ 15°C]{(CF_3CO)_2O} H_2NCH_2CH_2CH_2CH_2CH(NHCOCF_3)CO_2H\ \ 81\%\quad [439]$$

$$H_2NCH_2CH_2CH_2CH_2CH(NH_2)CO_2H \xrightarrow[NaOH,\ 20°C,\ 18\ hr]{CF_3COSC_2H_5} CF_3CONHCH_2CH_2CH_2CH_2CH(NH_2)CO_2H\ \ 75\%\quad [440]$$

Esters of fluorinated acids (fluoroacetic, trifluoroacetic acid) are converted by ammonia to the corresponding amides at low temperatures **[441]**.

(264) $$CF_3CO_2C_2H_5 \xrightarrow[0°C]{NH_3,\ Et_2O} CF_3CONH_2\ \ 99\%$$ **[441]**

*Acylations at Carbon*

Only very few examples of acylation at carbon are known; nevertheless, they seem to represent a feasible way to fluorinated $\beta$-diesters, $\beta$-ketoesters, or their derivatives **[442,443]**.

(265) $$C_6F_5COCl \xrightarrow{CH_2(CO_2C_2H_5)_2,\ Mg(OEt)_2} C_6F_5COCH(CO_2C_2H_5)_2\ \ 65\%\quad [442]$$

$$C_6F_5COCl \xrightarrow{CH_3COCH_2CO_2C_2H_5}$$ (tetrafluoro-benzopyranone ring, F, bearing C–$CO_2C_2H_5$ and =C–$CH_3$, ring O) 84% [443]

## Syntheses with Organometallic Compounds

Syntheses using organometallic compounds provide numerous means for the preparation of organic fluorine compounds, either by modifying the fluorinated substrates, or by introducing fluorinated groups by means of fluorine-containing organometallics. The latter type of reaction has recently been enriched by using perfluoro-organometallics.

Syntheses using organometallics lead most frequently to alcohols, aldehydes, ketones, and acids. Many deviations from the regular course of the reaction are encountered among fluorinated compounds, mainly owing to the different reactivities of fluorine when located in different positions in the organic molecule. Metals involved in fluorinated organometallics are magnesium, lithium, zinc, mercury, and copper.

### *The Grignard Syntheses*

Fluorine atoms may be present either in a substrate reacting with regular *Grignard reagents*, or in Grignard reagents themselves. In the latter case, only certain fluorinated compounds can be converted to fluorinated Grignard reagents.

*Grignard Reagents as Organic Substrate.* Fluorinated compounds were treated with all kinds of Grignard reagents derived from paraffins, olefins, acetylenes, or aromatics. The reactions usually give the expected products, with only a few exceptions noted. $\alpha$-Fluoroketones react normally at low temperatures, whereas at higher temperatures the fluorine atom is replaced by the carbanion **[444]**.

(266) [444]

$$C_6H_5COCH_2F \xrightarrow[0°C]{C_6H_5MgBr} C_6H_5C(CH_2F)(OH)C_6H_5 \quad 65\%$$

$$C_6H_5COCH_2F \xrightarrow[heat]{C_6H_5MgBr} C_6H_5COCH_2C_6H_5$$

Some Grignard reagents, especially those derived from secondary and tertiary alkyl halides, tend to reduce the organic carbonyl compound or a nitrile instead of undergoing the normal addition reaction **[445]**. Sometimes, the reduction is the main reaction.

(267) [445]

$$C_3F_7CN \xrightarrow[Et_2O,\ 0°C,\ regular\ addn.]{(CH_3)_2CHMgBr} C_3F_7CHO \quad 35\%$$

$$C_3F_7CN \xrightarrow[Et_2O,\ 0°C,\ reverse\ addn.]{(CH_3)_2CHMgBr} C_3F_7COCH(CH_3)_2 \quad 44\%$$

A series of interesting reactions results from the action of Grignard reagents upon fluoro-olefins and fluorohalo-olefins. Such reactions follow a $S_N2$ or $S_N2'$ mechanism in which the Grignard residue—a carbanion—is the attacking species. Consequently, similar rules are valid as in alkylation reactions (p. 113). In fluorohalo-olefins, fluorine is displaced preferentially to other halogens **[446,447]**.

$$CF_2{=}CCl_2 \longleftrightarrow \overset{\oplus}{C}F_2{-}\overset{\ominus}{\ddot{C}}Cl_2 + C_6H_5MgBr \longrightarrow [C_6H_5CF_2{-}\overset{\ominus}{C}Cl_2]MgBr \xrightarrow{-MgBrF} \underset{63\%}{C_6H_5CF{=}CCl_2} \tag{268}$$ **[446]**

(269) **[447]**

$$\text{cyclo-}(CF_2{-}CCl{=}CCl{-}CF_2) \xrightarrow[Et_2O,\ reflux]{C_2H_5MgBr} \text{cyclo-}(CF_2{-}CC_2H_5{=}CCl{-}CF_2)\ 75\% \xrightarrow{C_2H_5MgBr} 16\%\ \text{cyclo-}(CF_2{-}CC_2H_5{=}CC_2H_5{-}CF_2)$$

$$\text{cyclo-}(CF_2{-}CF{=}CF{-}CF_2) \xrightarrow[Et_2O,\ reflux]{C_2H_5MgBr} \text{cyclo-}(CF_2{-}CC_2H_5{=}CF{-}CF_2)\ 75\% \xrightarrow{C_2H_5MgBr} 75\%\ \text{cyclo-}(CF_2{-}CC_2H_5{=}CC_2H_5{-}CF_2)$$

Poor electron density in the nuclei of perfluoroaromatics provides for easy displacement of aromatic fluorine atoms by carbanions derived from Grignard reagents **[448]**.

$$C_6F_6 \xrightarrow[THF]{C_2H_5MgBr} \underset{56\%}{C_2H_5C_6F_5} + \underset{7\%}{HC_6F_5} \tag{270}$$ **[448]**

*Fluorinated Grignard Reagents.* Success in preparing fluorinated Grignard reagents depends largely on the mutual positions of the fluorine atom or atoms and the reactive halogen in the molecule of a fluorohalo compound. Organic fluorides do not react with magnesium to give Grignard reagents. Fluorohalides having fluorine atoms in vicinal positions to another halogen form olefins when treated with magnesium metal. Similarly, fluorine in the $\gamma$-position with respect to another halogen may give, by 1,3-elimination, a cyclopropane derivative. Under mild conditions, however, 3-chloro- and 3-bromo-1,1,1-trifluoropropane gave a Grignard reagent **[449,450]**.

(271)

$$CF_3CH_2CH_2Br \xrightarrow[Et_2O]{Mg} CF_3CH_2CH_2MgBr$$

[449] [450]

$CO(OCH_3)_2$ | $Et_2O$, 27% → $CF_3CH_2CH_2CO_2CH_3$

$CF_3CO_2CH_3$ | $Et_2O$, 70% [450] → $CF_3CH_2CH_2COCF_3$

$CF_3CO_2CH_3$ | THF, 23%, 53% → $(CF_3CH_2CH_2)_2CCF_3$ (OH)

The more distant fluorine atoms do not interfere with the formation of a fluorinated Grignard reagent. Nevertheless, some instances of halogen–metal interconversion were noticed in the preparation of Grignard reagents from $\alpha,\omega$-fluorohaloparaffins. Such a reaction is responsible for the occurrence of $\omega$-haloalkanecarboxylic acids and $\alpha,\omega$-dicarboxylic acids in addition to the expected $\omega$-fluoroalkanecarboxylic acids after the treatment of the reaction mixture with carbon dioxide **[451]**.

(272)

$$F(CH_2)_6X \xrightarrow[2.\ CO_2]{1.\ Mg,\ Et_2O} F(CH_2)_6CO_2H + X(CH_2)_6CO_2H + HO_2C(CH_2)_6CO_2H$$

| X = | $F(CH_2)_6CO_2H$ | $X(CH_2)_6CO_2H$ | $HO_2C(CH_2)_6CO_2H$ |
|---|---|---|---|
| Cl | 64% | 0% | 18% |
| Br | 31.5% | 22.5% | 12% |
| I | 0% | 23.4% | 0% |

**[451]**

In the aromatic series, a Grignard reagent was easily prepared from *m*-bromobenzotrifluoride **[452]**. If the fluorine atom is in *o*-position to bromine, the reaction with magnesium results in the formation of dehydrobenzene (benzyne) (p. 127).

(273)

$$m\text{-}CF_3C_6H_4Br \xrightarrow[Et_2O]{Mg} m\text{-}CF_3C_6H_4MgBr \xrightarrow{C_6H_5NMeCHO} m\text{-}CF_3C_6H_4CHO$$

54–55% 59% **[452]**

*Perfluorinated Grignard Reagents.* It is not without some surprise that even perfluorinated Grignard reagents can be prepared (Henne, 1951). Both theoretically and according to experience, a perfluoroalkyl halide should eliminate fluorine and the neighboring halogen and give an olefin. Nevertheless, it is possible to prepare perfluorinated Grignard reagents under special conditions. It is essential to use pure chemicals and to cool the reaction mixture well below zero as soon as the reaction has started. Perfluoroalkylmagnesium halides are very unstable at room temperature, and are best prepared at temperatures of −80 to −30°C. Tetrahydrofuran is preferable to diethyl ether. Trifluoromethylmagnesium halides are the most difficult to prepare since the reactivity of trifluoromethyl iodide is about 0.001 of that of perfluoropropyl iodide **[453,454]**.

(274)

$C_3F_7MgCl \xrightarrow[-80°C]{CO_2} C_3F_7CO_2H$ 51%

$C_3F_7MgCl \xrightarrow{(CH_3)_2CO} (CH_3)_2C(OH)C_3F_7$ 10%

$C_3F_7MgCl \xrightarrow[-50° \text{ to } -20°C]{C_3F_7CHO} C_3F_7CH(OH)C_3F_7$ 16%

$C_3F_7MgCl \xrightarrow[-80° \text{ to } -30°C]{HCO_2C_2H_5} C_3F_7CH(OH)_2$ 24%

$C_3F_7MgCl \xrightarrow[-50° \text{ to } -30°C]{C_3F_7CO_2C_2H_5} C_3F_7COC_3F_7$ 20%

[453, 454]

In addition to the regular direct preparation of perfluoroalkyl Grignard reagents, an indirect route can be used. This is based on halogen–metal interchange between a perfluoroalkyl halide and a common Grignard reagent **[455]**.

(275)

$C_6H_5MgBr \xrightarrow[Et_2O,\ -80°C]{C_2F_5I,\ CH_3COCH_3} (CH_3)_2C(OH)C_2F_5$ 38%

$C_6H_5MgBr \xrightarrow[Et_2O,\ 0°C,\ 12\ hr]{C_3F_7I,\ CH_3COCH_3} (CH_3)_2C(OH)C_3F_7$ 65%

[455]

*Perfluorovinylmagnesium compounds* were also prepared, both by the direct reaction of perfluorovinyl halides with magnesium **[456]** and by halogen–metal interchange from perfluorovinyl iodide and phenylmagnesium bromide **[457]**.

(276)

$CF_2{=}CFI \xrightarrow[-20°C]{Mg,\ Et_2O} CF_2{=}CFMgI \xrightarrow{H_2SO_4} CF_2{=}CHF$ 69% [456]

$CF_2{=}CFMgI \xrightarrow{CO_2}$ 37.7%

$CF_2{=}CFI \xrightarrow{C_6H_5MgBr} CF_2{=}CFMgBr \xrightarrow{CO_2} CF_2{=}CFCO_2H$ 32% [457]

An interesting rearrangement takes place in the reaction of perfluorovinylmagnesium halides with aldehydes or ketones, the primary products, carbinols, being converted to $\alpha,\beta$-unsaturated acids **[458]**.

(277)

$CF_2{=}CFBr \xrightarrow[0°C,\ 1\ hr,\ 20°C,\ 1\ hr]{Mg,\ THF} CF_2{=}CFMgBr \xrightarrow{CF_3COCH_3} CF_2{=}CFC(CH_3)(CF_3)OH$ 29%

$\xrightarrow{heat} [HOCF_2CF{=}C(CH_3)CF_3 \longrightarrow FCOCF{=}C(CH_3)CF_3] \longrightarrow HO_2CCF{=}C(CH_3)CF_3$ 11%

[458]

*Acetylenic perfluorinated Grignard reagents* were prepared by hydrogen–metal interchange between fluorinated acetylenes having acetylenic hydrogen, such as 3,3,3-trifluoropropyne, and a conventional Grignard reagent **[458a]**.

Preparation of *perfluoroaromatic Grignard reagents* is not nearly as demanding as that of the aliphatic reagents. No special precautions are needed for the preparation of perfluorophenylmagnesium halides. Addition reactions of these reagents to fluorinated or nonfluorinated carbonyl compounds or to carbon dioxide provide means for introducing perfluoroaryl groups into organic compounds **[459,460]**.

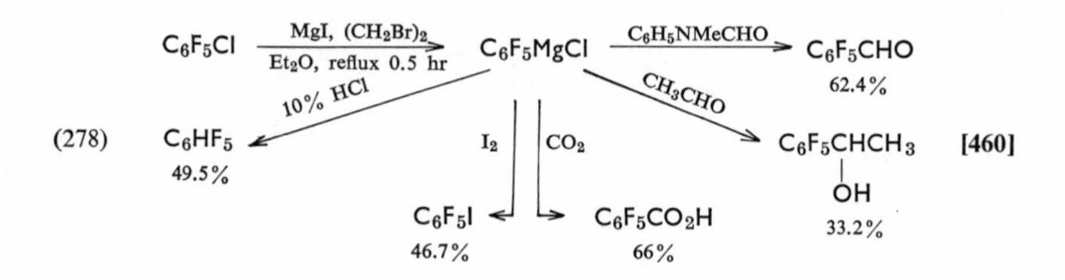


Since hydrogen in pentafluorobenzene is acidic in nature, pentafluorobenzene may be converted to pentafluorophenyl magnesium halides by hydrogen–metal interchange when treated with Grignard reagents. This reaction gives excellent yields **[461]**. Examples of synthetic potentialities of perfluorophenylmagnesium bromide are shown below **[462,463]**.

(279)

- $C_6F_5Br \xrightarrow[20\text{–}26°C,\ 16\ hr]{Mg,\ Et_2O} C_6F_5MgBr$
- $C_6HF_5 \xrightarrow[5\ hr]{C_2H_5MgBr,\ THF} C_6F_5MgBr$ (86%)
- $C_6F_5MgBr \xrightarrow{CO_2} C_6F_5CO_2H$ (85%) [461]
- $C_6F_5MgBr \xrightarrow{IC{\equiv}CI} C_6F_5C{\equiv}CC_6F_5$ [463]
- $C_6F_5MgBr \xrightarrow{CClF_2CHFI} C_6F_5I$ (43%) [462]

The preparation of a perfluorinated Grignard reagent is possible even in the pyridine series **[464]**.

(280) Tetrafluoro-4-bromopyridine $\xrightarrow[-20°C \text{ to } 0°C]{Mg, THF}$ 2,3,5,6-tetrafluoro-4-pyridylmagnesium bromide; with $CO_2$ → 2,3,5,6-tetrafluoroisonicotinic acid ($CO_2H$) 59%; with $C_2H_5COCH_3$ (−10°C, 0.5 hr; 20°C, 1.5 hr) → $C_2H_5C(OH)CH_3$-substituted tetrafluoropyridine; with $C_5F_5N$ → octafluoro-4,4′-bipyridyl 68% [464]

*Organolithium Compounds*

Next to Grignard reagents, organolithium compounds have been best explored. Their preparation and application parallel those of the Grignard reagents. However, in some instances, differences result from the higher reactivity of organolithium compounds.

*Organolithium Compounds as Organic Substrate.* Like Grignard reagents, lithium compounds react with fluorinated derivatives according to a general scheme of nucleophilic addition of carbanions. Such reactions are good preparative methods for the syntheses of fluorinated alcohols from fluorinated carbonyl compounds, and of fluorinated ketones from fluorinated acids or their derivatives **[465]**.

(281) $$CF_3CO_2Li + C_4H_9Li \xrightarrow[\text{reflux 1 hr}]{Et_2O} CF_3COC_4H_9 \quad 61\%$$ [465]

In fluorinated olefins and halo-olefins, carbanions derived from lithium compounds displace fluorine or other halogen atoms. The outcome of the reaction depends on the mechanism ($S_N2$ or $S_N2'$) and on the leaving ability of the halogen **[466,467]**.

(282) $$CH_3C{\equiv}CLi + CF_2{=}CClF \xrightarrow{THF} (CH_3C{\equiv}C)FC{=}CFCl \;(14.6\%) + (CH_3C{\equiv}C)FC{=}CClF \;(58.4\%)$$ [466]

(283)

$$C_6H_5C\overset{}{=}CH-CF_2-CFCl \xrightarrow[\text{Et}_2\text{O},\ 0°\text{C, 2 hr}]{C_6H_5Li} C_6H_5C(=CF)-CF_2-CH(C_6H_5)\ (73\%) \xrightarrow[\text{Et}_2\text{O}]{C_6H_5Li} C_6H_5C(=C(C_6H_5))-CF_2-CH(C_6H_5)\ (38\%)$$

2 $C_6H_5Li$ [467]

Fluorine in perfluoroaromatics is relatively easily displaced by carbanions in reactions with organolithium compounds **[468,469]**.

(284)

$C_6F_6$ + LiCH=CHCH$_3$ → $C_6F_5$–CH=CHCH$_3$ 87% *cis+trans* [468]

$C_6F_6$ + 2 LiCH=CHCH$_3$ → CH$_3$CH=CH–$C_6F_4$–CH=CHCH$_3$ 82% *cis, cis* [468]

$C_6F_6$ + $C_6H_5C{\equiv}CLi$ → $C_6H_5C{\equiv}C$–$C_6F_4$–$C{\equiv}C$–$C_6H_5$ 62.5% [469]

The hydrogen atom in pentafluorobenzene is so acidic that it reacts with organolithium compounds in a metathetical reaction and gives pentafluorophenyllithium **[470,471]**. This reaction is an alternative route to fluorinated organolithium compounds, which will be dealt with in the next paragraph.

(285) [470] [471]

$$C_6HF_5 \xrightarrow[-60° \text{ to } -50°C]{C_4H_9Li,\ THF} C_6F_5Li$$

$$C_6F_5Li \xrightarrow[-60° \text{ to } -45°C]{S} C_6F_5SH\ (45.7\%)$$

$$C_6F_5Li \xrightarrow[-65°C]{B(OCH_3)_3,\ THF} C_6F_5B(OCH_3)_2\ (85\%) \xrightarrow{H_2O_2} C_6F_5OH\ (39\%)$$

*Fluorinated Organolithium Compounds*. Similar limitations to those in the preparation of fluorinated Grignard reagents also hold true for the preparation of fluorinated organolithium compounds: elimination of fluorine and halogen results from the action of lithium on compounds having fluorine and a halogen in 1,2- or 1,3-positions.

Perfluoro-olefins and polyfluorohalo-olefins containing hydrogen can be converted to lithium derivatives by hydrogen–metal interchange **[472,476]**.

$$(286)\quad CClF{=}CHCl \xrightarrow{C_4H_9Li} CClF{=}CClLi \xrightarrow{(CH_3)_2CO} CClF{=}CClC(CH_3)_2OH\ \ 60\% \quad [476]$$

Surprisingly, even in monofluorobenzene, hydrogen atom in the *ortho* position to fluorine is replaced by lithium when the compound is treated with butyllithium **[473]**. Thus, not only *o*-bromofluorobenzene, but fluorobenzene itself is suitable starting material for the preparation of *o*-fluorophenyllithium. This compound is stable only at low temperatures. At temperatures approaching 0°C, lithium fluoride is eliminated and dehydrobenzene, or benzyne, is formed **[474]**. Preparation of *o*-trifluoromethylphenyllithium by treatment of benzotrifluoride with butyllithium was mentioned previously (p. 100) **[334]**.

(287)

Fluorobenzene **[473]** —($C_4H_9Li$, THF, −50° to −60°C)→ *o*-fluorophenyllithium ←($C_4H_9Li$, $Et_2O$, <−50°C)— *o*-bromofluorobenzene **[474]**

*o*-fluorophenyllithium —($CO_2$, −50°C)→ *o*-fluorobenzoic acid ($CO_2H$), 60%

*o*-fluorophenyllithium —(furan, −10°C)→ 1,4-dihydro-1,4-epoxynaphthalene, 67%

*Perfluorinated Organolithium Compounds. Perfluoroalkyllithium compounds* can be prepared from perfluoroalkyl halides at low temperatures. At temperatures above 0°C, the organometallics decompose to perfluoroolefins **[475]**.

(288) **[475]**

$$C_3F_7I \xrightarrow[-74°C]{CH_3Li,\ Et_2O} C_3F_7Li \xrightarrow{20°C} CF_3CF{=}CF_2 \quad 93\%$$

$$C_3F_7Li \xrightarrow[\text{reflux}]{H_2SO_4} CF_3CF_2CHF_2 \quad 64\%$$

$$C_3F_7Li \xrightarrow[-42°C]{C_2H_5CHO} CF_3CF_2CF_2CH(OH)CH_2CH_3 \quad 50\%$$

*Perfluorovinyllithium* was prepared both by halogen–metal and hydrogen–metal interchange. When treated with carbonyl compounds, perfluorovinyllithium gives perfluorovinylcarbinols which rearrange and hydrolyze to $\alpha,\beta$-unsaturated acids **[472]**.

(289) **[472]**

$$CF_2{=}CFBr \text{ or } CF_2{=}CHF \xrightarrow{C_4H_9Li} CF_2{=}CFLi \xrightarrow{(CH_3)_2CO} CF_2{=}CFC(CH_3)_2OH \longrightarrow HOCOCF{=}C(CH_3)_2 \quad 30\%$$

Also, *perfluorophenyllithium* can be prepared both from pentafluorobromobenzene and pentafluorobenzene **[477,478]**. As in the case of *o*-bromofluorobenzene (p. 127), lithium fluoride is split out at room temperature and a dehydrobenzene (benzyne) derivative is formed. The synthetic versatility and usefulness of pentafluorophenyllithium are shown in the following equations:

(290) **[477]**

$$C_6F_5Br \xrightarrow[-78°C,\ 5\ min]{C_4H_9Li} C_6F_5Li \xleftarrow[Et_2O]{C_4H_9Li} C_6HF_5$$

$$C_6F_5Li \xrightarrow[15°C]{H_2O} C_6HF_5$$

$$C_6F_5Li \xrightarrow{CO_2} C_6F_5CO_2H \quad 68\%$$

$$C_6F_5Li \xrightarrow[15°C]{\text{furan}} \text{tetrafluoro-1,4-dihydro-1,4-epoxynaphthalene} \quad 48\%$$

$$C_6F_5Li \xrightarrow{HCONHMe} C_6F_5CHO \quad 40\%$$

$$C_6F_5Li \xrightarrow{HCO_2C_2H_5} C_6F_5CHO \quad 18.5\%$$

## *Organozinc Compounds*

*Perfluoropropylzinc iodide* was prepared by direct synthesis from zinc and perfluoropropyl iodide. It is much more stable than the corresponding organomagnesium compound or organolithium compound, and forms perfluoropropene only on heating **[479]**.

(291) **[479]**

$$C_3F_7I \xrightarrow[0°C]{Zn,\ dioxane} C_3F_7ZnI \ (50\text{–}60\%) \xrightarrow{180\text{–}200°C} C_3F_6 \quad 58\%$$

$$C_3F_7ZnI \xrightarrow{Cl_2} C_3F_7Cl \quad 89\%$$

$$C_3F_7ZnI \xrightarrow[100°C]{H_2O} C_3HF_7 \quad 96\%$$

$$C_3F_7ZnI \xrightarrow{CH_3COCl} C_3F_7COCH_3 \quad 18\%$$

Organozinc compounds are intermediates in the *Reformatsky* synthesis of fluorinated $\beta$-hydroxy-esters from $\alpha$-bromoesters and fluorinated ketones **[480]**.

(292) **[480]**

$$(CH_3)(CH_2F)CO + BrCH_2CO_2C_2H_5 \xrightarrow[50\text{–}60°C,\ 3\ hr]{Zn,\ C_6H_6} (CH_3)(CH_2F)C(OH)CH_2CO_2C_2H_5 \quad 40\%$$

## *Organomercury Compounds*

Fluorinated organomercury compounds are numerous and their synthetic applications are versatile. Some of them are shown below **[481,482]**:

(293)

$$C_6F_5CO_2H \xrightarrow[AcOH]{Hg(OAc)_2} (C_6F_5CO_2)_2Hg \ (87\%) \xrightarrow[65\ min]{210\text{–}212°C} (C_6F_5)_2Hg \ (56\%)$$

$$(C_6F_5)_2Hg \xrightarrow[250°C,\ 7\ days]{S} (C_6F_5)_2S \ (82\%) \xleftarrow[6\ days]{S,\ 250°C} C_6F_5HgCl$$

*Organocopper Compounds*

Fluorinated copper compounds are the probable intermediates in the Wurtz–Fittig synthesis of polyfluorinated aromatics **[483]**:

(294) $I(CF_2)_3I$ + I—⟨F⟩—$CO_2CH_3$ $\xrightarrow[120°C]{Cu}$ $CH_3O_2C$—⟨F⟩—$(CF_2)_3$—⟨F⟩—$CO_2CH_3$ 85% + $CH_3O_2C$—⟨F⟩—⟨F⟩—$CO_2CH_3$ 15%

*Organometalloids*

Fluorinated compounds of silicon, phosphorus, arsenic, and antimony are generally available through the Grignard synthesis from the corresponding halides **[484]**. Sometimes, direct synthesis from the elements and fluorinated halides is feasible **[485]**.

(295) $CF_2{=}CFMgI$ $\xrightarrow[-18\text{ to }-10°C]{PCl_3}$ $(CF_2{=}CF)_3P$ 35.4%; $\xrightarrow[-18\text{ to }-10°C]{SbCl_3}$ $(CF_2{=}CF)_3Sb$ 41%; $\xrightarrow[-18\text{ to }-10°C]{AsCl_3}$ $(CF_2{=}CF)_3As$ 40% **[484]**

(296) $CF_3I$ $\xrightarrow[220°C]{P}$ $CF_3PI_2 + (CF_3)_2PI + (CF_3)_3P$; $CF_3I$ $\xrightarrow[220°C]{As}$ $CF_3AsI_2 + (CF_3)_2AsI + (CF_3)_3As$ **[485]**

## Base-Catalyzed Condensations

In this section dealing with base-catalyzed condensations, reactions will be described which involve addition of carbanions derived from nitroparaffins, aldehydes, ketones, esters, and nitriles across carbonyl bonds in aldehydes, ketones, and esters, or to a $\beta$-position in the $\alpha,\beta$-unsaturated versions of these compounds. The main representatives of these reactions are *aldol* condensations and *Claisen* condensations. Also included will be *cyanohydrin* synthesis, *Michael* addition, and *Wittig* synthesis.

Nitroparaffins condense with fluorinated aldehydes and ketones and form $\beta$-hydroxy derivatives which only unwillingly lose water when heated with strong dehydrating reagents **[486,487]**:

(297) [486] $CH_3NO_2$ $\xrightarrow[K_2CO_3,\ 50\text{–}60°C]{C_3F_7CH(OH)_2}$ $C_3F_7CH(OH)CH_2NO_2$ 75% $\xrightarrow[\text{reflux 1.5 hr}]{P_2O_5}$ $C_3F_7CH{=}CHNO_2$ 68%

[487] $CH_3NO_2$ $\xrightarrow[K_2CO_3,\ 120°C]{CF_3COCF_3}$ $(CF_3)_2C(OH)CH_2NO_2$ 78.4%

*Cyanohydrin synthesis* is a very useful means for the synthesis of fluorinated α-hydroxy-acids **[488]**:

(298) $$CF_3COCH_3 \xrightarrow[2.\ H_2SO_4,\ 0°C]{1.\ NaCN,\ 0°C} CF_3C(CN)(OH)CH_3 \quad 62.5\%$$ **[488]**

Fluorinated ketones having hydrogen atoms on α-carbons undergo regular *aldol condensation*. The aldols are stable, and dehydrate only with difficulty **[489]**:

(299) $$CF_3COCH_3 \xrightarrow{C_2H_5ONa,\ Et_2O,\ <0°C} 59\%;\quad \xrightarrow{Ca(NH_2)_2,\ NH_3,\ Et_2O} 38.7\% \quad CF_3C(CH_3)(OH)CH_2COCF_3$$ **[489]** **[250]**

In this connection, some of the syntheses using phosphorus-containing intermediates will be described. Starting materials are aldehydes, or ketones, trialkyl- or triarylphosphines, and α-halogenocarbonyl compounds, α-halogeno-esters, or other compounds with sufficiently reactive halogen atoms. Reactions of this type are referred to as *Wittig* reactions **[490–492]**. Their applications are as shown:

(300) $$(CF_3)(R)CO + (C_6H_5)_3P{=}CHCOR' \xrightarrow[reflux\ 3\ hr]{C_6H_6} (CF_3)(R)C{=}CHCOR'$$ **[490]**

| R = | $CH_3$ | $CH_3$ | $C_6H_5$ |
|---|---|---|---|
| R′ = | $CH_3$ | $C_6H_5$ | $CH_3$ |
| % | 58 | 70 | 77 |

(301) $$(C_6H_5)_3P + ClCF_2CO_2Na \longrightarrow (C_6H_5)_3\overset{\oplus}{P}{-}CF_2{-}CO{-}\overset{\ominus}{O} \longrightarrow (C_6H_5)_3P{=}CF_2$$

$$(C_6H_5)_3P{=}CF_2 + OC(CF_3)R \longrightarrow (C_6H_5)_3P{=}CF_2 \cdots O{=}C(CF_3)R \xrightarrow[100\text{–}105°C,\ 20\ hr]{(OCH_2OMe)_2} (C_6H_5)_3PO + CF_2{=}C(CF_3)R$$

| R = | $C_4H_9$ | $C_6H_{11}$ | $C_6H_5$ | $C_6H_5CH_2$ |
|---|---|---|---|---|
| % | 59 | 65 | 68 | 61 |

**[491]**

(302)

$$(C_6H_5)_3P + C_6F_5CH_2Br \xrightarrow[reflux,\ 5\ hr]{C_6H_6} [(C_6H_5)_3\overset{\oplus}{P}CH_2C_6F_5]\overset{\ominus}{Br}\ (97.5\%) \xrightarrow[C_6H_6,\ 15°C,\ 0.5\ hr]{C_4H_9Li} (C_6H_5)_3P{=}CHC_6F_5$$

$$\xrightarrow[20°C,\ 1.75\ hr]{C_6F_5COCl} (C_6H_5)_3P{=}C(C_6F_5)C(=O)C_6F_5\ (74\%) \xrightarrow[1\ hr]{310°C,\ 10\ min} (C_6H_5)_3PO + C_6F_5C{\equiv}CC_6F_5\ (95\%)$$ **[492]**

Condensation of aldehydes or ketones with $\alpha$-fluorinated esters leads to $\alpha$-fluoro-$\beta$-hydroxyesters, and ultimately to $\alpha$-fluoro-$\alpha,\beta$-unsaturated esters or acids **[493,494]**.

(303)

$$RCHO + CHF(COCO_2C_2H_5)CO_2C_2H_5 \xrightarrow[\text{xylene, reflux 15 min}]{\text{NaH, EtOH}} RCH(OH)CF(COCO_2C_2H_5)CO_2C_2H_5$$

$$RCH{=}CFCO_2H \ (66\text{–}70\%) \xleftarrow[<50°C]{\text{KOH, EtOH}} RCH{=}CFCO_2C_2H_5 \ (45\text{–}75\%) \quad [493]$$

*Knoevenagel-type reactions* were run with fluorinated ketones and malonic acid or its derivatives **[495,496]**, *Erlenmeyer* amino acid synthesis with fluorinated aldehydes and azlactones **[497]**.

(304)

$$CF_3COCF_3 \xrightarrow[50\text{–}100°C]{CH_2(CO_2C_2H_5)_2} (CF_3)_2C(OH)CH(CO_2C_2H_5)_2 \ (78\%) \quad [495]$$

$$CF_3COCF_3 \xrightarrow[\text{ZnCl}_2,\ 80°C,\ 2\ \text{hr}]{CH_2(CN)_2} (CF_3)_2C(OH){-}CH(CN)_2 \ (94\%) \xrightarrow[\text{distn.}]{P_2O_5} (CF_3)_2C{=}C(CN)_2 \ (50\%) \quad [496]$$

(305)

$$C_6F_5CHO + \text{2-phenyl-5-oxazolone (azlactone)} \xrightarrow[K_2CO_3]{(CH_3CO)_2O} C_6F_5CH{=}C\text{(azlactone)} \ (93\%) \quad [497]$$

$$\xrightarrow[\text{reflux 3.5–4 hr}]{\text{NaOH}} C_6F_5CH{=}C(NHCOC_6H_5)CO_2H \ (82\%) \xrightarrow[0°C,\ 5\ \text{hr}]{\text{Zn, AcOH}} C_6F_5CH_2CH(NHCOC_6H_5)CO_2H \ (88\%) \xrightarrow[\text{reflux 5 hr}]{\text{HBr, AcOH}} C_6F_5CH_2CH(NH_2)CO_2H \ (94\%)$$

*Claisen condensation* of esters with ketones and of esters with esters are some of the most important synthetic routes to fluorinated products or intermediates. The former reaction leads to fluorinated $\beta$-diketones **[498, 499]**, the latter to fluorinated $\beta$-ketoesters **[500–505]**.

(306)

$$CF_3CO_2C_2H_5 \xrightarrow[\text{EtONa, Et}_2\text{O, reflux 50 hr}]{CH_3COCH_3} CF_3COCH_2COCH_3 \ (70\%) \quad [498]$$

$$CF_3CO_2C_2H_5 \xrightarrow[\text{EtONa, Et}_2\text{O, reflux 50 hr}]{CH_3COCF_3} CF_3COCH_2COCF_3 \ (75\%) \quad [498]$$

$$CF_3CO_2C_2H_5 \xrightarrow{\text{2-furyl-COCH}_3} \text{2-furyl-COCH}_2COCF_3 \ (92.5\%)$$

$$CF_3CO_2C_2H_5 \xrightarrow[\text{EtONa, 90°C, 12 hr}]{C_6H_5COCH_3} C_6H_5COCH_2COCF_3 \ (70\%) \quad [499]$$

$$CF_3CO_2C_2H_5 \xrightarrow{\text{2-thienyl-COCH}_3} \text{2-thienyl-COCH}_2COCF_3 \ (88\%)$$

**Table 46. Claisen Condensation of Fluorinated Acetates and Ketonic Fission of the Fluorinated Acetoacetates [505]**

| Fluorinated acetoacetate | Reaction temperature, °C | Yield, % | Fluorinated acetone | Yield, % |
|---|---|---|---|---|
| $CF_3COCF_2COOC_2H_5$ | 75 | 20 | $CF_3COCHF_2$ | 35 |
| $CHF_2COCF_2COOC_2H_5$ | 70 | 81 | $CHF_2COCHF_2$ | 62 |
| $CF_3COCHFCOOC_2H_5$ | 50 | 86 | $CF_3COCH_2F$ | 67 |
| $CF_3COCH_2COOC_2H_5$ | 50 | 84 | — | — |
| $CHF_2COCHFCOOC_2H_5$ | 30 | 68 | $CHF_2COCH_2F$ | 65 |
| $CHF_2COCH_2COOC_2H_5$ | 40 | 83 | — | — |
| $CH_2FCOCHFCOOC_2H_5$ | 40 | 69 | $CH_2FCOCH_2F$ | 74 |

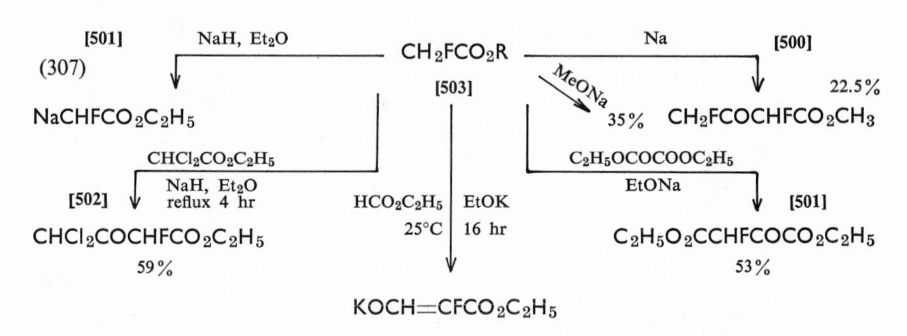


(308)

$$CF_3COCH_2CO_2C_2H_5 \; 55\% \xleftarrow[\text{Na, Et}_2\text{O, reflux 14 hr}]{CF_3CO_2C_2H_5} CH_3CO_2C_2H_5 \xrightarrow{C_3F_7CO_2C_2H_5} C_3F_7COCH_2CO_2C_2H_5 \; 32.5\%$$ [504]

Ketonic fission of fluorinated $\beta$-ketoesters is an easy way to fluorinated ketones (Table 46) **[505]**.

*Michael addition* of fluorinated esters to acrylates, or esters to fluorinated acrylates, is exemplified in the synthesis of $\gamma$-fluoroglutamic acid **[506]**:

(309)

$$CH_2{=}C(NHCOCH_3)CO_2C_2H_5 + CHF(CO_2C_2H_5)_2 \xrightarrow[\text{EtOH}]{\text{Na}} (C_2H_5O_2C)_2CFCH_2CH(NHCOCH_3)CO_2C_2H_5$$

$$(C_2H_5O_2C)_2CFCH_2CH(NHCOCH_3)CO_2C_2H_5 \xrightarrow{\text{HCl}} HO_2CCHFCH_2CH(NH_2)CO_2H \; 56\%$$

$$CH_2{=}CFCO_2CH_3 + CH(NHCOCH_3)(CO_2C_2H_5)_2 \longrightarrow CH_3O_2CCHFCH_2C(NHCOCH_3)(CO_2C_2H_5)_2 \xrightarrow{\text{HCl}} HO_2CCHFCH_2CH(NH_2)CO_2H$$ [506]

## ADDITIONS

Addition reactions are very abundant in the chemistry of organic fluorine compounds, and many of them have already been discussed under different labels: electrophilic additions leading to halogen derivatives (p. 99, 101), and additions of alkyl halides to fluoro-olefins (p. 105). Also, reactions involving addition–elimination have been mentioned in sections on alkylations (pp. 112, 114) and on syntheses with organometallics (p. 121). In the following paragraphs, the remaining nucleophilic additions to fluoro-olefins, and free-radical additions to olefins and acetylenes will be dealt with.

### Nucleophilic Additions to Fluorinated Olefins

The inductive effect of fluorine in polyfluoro-olefins accounts for the high reactivity of the double bond toward nucleophiles, while the strong mesomeric effect of fluorine is responsible for the orientation of addition. Consequently, a nucleophile always combines preferentially with the carbon atom holding two fluorine atoms.

(310) $$CClF{=}CF_2 \longleftrightarrow \overset{\ominus}{\underset{\cdot\cdot}{C}}ClF{-}\overset{\oplus}{C}F_2 \xrightarrow{\overset{\partial+\ \partial-}{H{-}Y}} CHClFCF_2Y$$

*Additions of Alcohols and Phenols*

Fluorohalo-olefins and perfluoro-olefins add alcohols and phenols in the presence of alkaline catalysts **[507–509]**. This reaction is of much practical importance, since $\alpha,\alpha$-difluoroethers thus formed easily hydrolyze to fluorinated esters or acids (p. 110).

(311) $$CClF{=}CF_2 \xrightarrow[2.5\ hr]{C_2H_5OH,\ C_2H_5ONa} \underset{88\text{–}92\%}{CHClFCF_2OC_2H_5}$$ **[507-509]**

Some fluoro-olefins, especially the branched ones and the fluorocyclo-olefins, tend to form unsaturated ethers by subsequent elimination of hydrogen fluoride. This reaction prevails with higher alcohols **[510]**.

(312) $$(CF_3)_2C{=}CF_2 \xrightarrow[20\text{–}40°C]{ROH} (CF_3)_2CHCF_2OR + (CF_3)_2C{=}CFOR$$ **[510]**

| R = | $(CF_3)_2CHCF_2OR$ | $(CF_3)_2C{=}CFOR$ |
|---|---|---|
| $CH_3$ | 70% | 0% |
| $(CH_3)_2CH$ | 35% | 26% |

*Additions of Mercaptans and Thiophenols*

Addition of mercaptans and thiophenols follows the same rules as that of their oxygen-containing analogs. Elimination of hydrogen fluoride is here even more pronounced than with alcohols and phenols, especially in the perfluorocyclobutene series **[511]**.

(313)
$$\underset{CF_2-CF_2}{CF=CF} \xrightarrow[65^\circ C,\ 30\ min]{C_4H_9SH,\ Triton\ B} \underset{CF_2-CF_2}{C_4H_9SCH-CFSC_4H_9}\quad 30\%$$
[511]

*Additions of Ammonia and Amines*

Primary addition products of fluorinated olefins and ammonia readily undergo consecutive reactions, either elimination of hydrogen fluoride leading to nitriles, or hydrolysis leading to amides **[512]**.

(314)
$$(CF_3)_2C=CF_2 \xrightarrow[Et_2O,\ -60\rightarrow 0^\circ C]{NH_3} [(CF_3)_2CHCF_2NH_2] \begin{cases} \xrightarrow{-HF} (CF_3)_2CHCN\quad 21\% \\ \xrightarrow{H_2O} (CF_3)_2CHCONH_2\quad 13\% \end{cases}$$
[512]

Primary amines and perfluoro-olefins give enamines or Schiff's bases, while secondary amines give $\alpha,\alpha$-difluoroamines, which sometimes, especially in the fluorocyclobutene series, eliminate hydrogen fluoride and give also enamines **[513]**.

(315)
$$\underset{CF_2-CF_2}{CF=CF} \xrightarrow[15\text{–}25^\circ C,\ 4\ hr]{4\ C_4H_9NH_2} \underset{CF_2-CF_2}{C_4H_9NHC=CNHC_4H_9} \longrightarrow \underset{CF-CF_2}{C_4H_9NHC-C=NC_4H_9} \longrightarrow \underset{HCF-CF_2}{C_4H_9N=C-C=NC_4H_9}\ \ 72\%$$
$$\underset{CF_2-CF_2}{CF=CF} \xrightarrow[15\text{–}25^\circ C,\ 4\ hr]{NH(C_2H_5)_2} \underset{CF_2-CF_2}{(C_2H_5)_2NC=CF}\ \ 59\%$$
[513]

The preparation of chlorotrifluorotriethylamine from chlorotrifluoroethylene and diethylamine was mentioned previously, since the product reacts readily with hydroxyl-containing compounds and converts acids to acyl fluorides and alcohols to alkyl fluorides. The compound itself is converted to N,N-diethyl chlorofluoroacetamide [Equation (77)].

## Free-Radical-Type Additions

In contrast to nucleophilic additions, free-radical-type additions are preceded by homolytic fission of bonds. This fission is achieved by heat, irradiation, or organic free-radical sources such as dibenzoyl peroxide. The reactions can be subdivided into linear additions and cycloadditions, which in turn can be distinguished into three-center or four-center reactions.

*Linear Additions*

The most common additions of the linear type are the reactions of haloparaffins or fluorohaloparaffins with *olefins* or *fluorinated olefins*. The reaction has a free-radical chain mechanism. Consequently, the 1:1 adducts are accompanied by varying amounts of telomers containing multiple units of the starting olefin. The ratio of products depends largely on the ratio of the reactants **[514]**.

$$\text{(316)} \quad CH_2{=}CF_2 \xrightarrow[\text{80°C, 6 hr, sealed}]{CF_2Br_2,\ Bz_2O_2} \underset{33\text{–}35\%}{CF_2BrCH_2CF_2Br} + \underset{23\text{–}29\%}{CF_2Br(CH_2CF_2)_2Br} \quad [514]$$

Although the free-radical additions are usually unidirectional, two isomeric products were isolated from some reactions **[515]**.

$$\text{(317)} \quad CHF{=}CF_2 \xrightarrow[h\nu]{CF_3I} \underset{68\%}{CF_3CHFCF_2I} + \underset{17\%}{CF_3CF_2CHFI} \quad [515]$$

The attacking free radical generated from fluorohaloparaffins by homolytic fission is the carbon-containing residue. Its combination with a carbon atom of a double bond in the olefin is probably accomplished in such a way as to form a more stable intermediate radical. Reactivity of individual fluoro-olefins toward such additions decreases in the series **[516]**

Decreasing reactivity of olefins

$$\text{(318)} \quad CH_2{=}CF_2 > CF_2{=}CF_2 > CF_2{=}CHCl > CF_2{=}CFCl > CF_2{=}CFCF_3 > CF_3CF{=}CFCF_3 > \begin{matrix} CF{=}CF \\ | \quad\ \ | \\ CF_2{-}CF_2 \end{matrix} > CF_2\langle\begin{matrix} CF{=}CF \\ CF_2{-}CF_2 \end{matrix}\rangle CF_2 \quad [516]$$

The ease of attack by a free radical, whether determined by steric or thermodynamic factors, decreases in the series

Increasing stability of free radicals

$$\text{(319)} \quad \dot{C}H_2 < \dot{C}HF < \dot{C}F_2 < \begin{matrix} \dot{C}HCl \\ \dot{C}FCl \end{matrix} < \begin{matrix} \dot{C}HCH_3 \\ \dot{C}HCF_3 \end{matrix} < \dot{C}F_2 < \dot{C}FCF_3$$

Table 47 lists numerous examples of free-radical additions to olefins and fluorinated olefins. These reactions are of great importance from the synthetic point of view. The products can be used for the preparation of organometallics, or they can be converted to olefins by elimination of hydrogen halides. The olefins thus formed may be converted to acetylenes, or oxidized to acids or ketones, etc.

Not only fluorohaloparaffins, but also fluorohalo-olefins can add across double bonds in olefins and fluoro-olefins in the way that the organic radical joins one carbon and the halogen the other carbon of the double bond. Fluorohalo-olefins are thus formed as final products **[517]**.

**Table 47. Addition of Haloparaffins to Olefins and Halo-olefin[a]**

| Olefin[b] | $CHCl_3$ | $CCl_4$ | $CCl_3Br$ | $CCl_3I$ | $CF_3I$ | $CF_2Br_2$ | $CHFCl_2$ | $CF_2ClBr$ | $C_2F_5I$ | $C_3F_7I$ | $CF_2BrCFClBr$ |
|---|---|---|---|---|---|---|---|---|---|---|---|
| $CH_2{=}CH_2$ | — | — | — | — | [522] [524] | [536] | — | — | — | — | — |
| $CH_2{=}CHF$ | — | — | — | — | — | [538] | — | — | — | — | — |
| $CH_2{=}CF_2$ | — | — | — | — | [528] | — | — | [537] | [529] | — | [535] |
| $CHF{=}CF_2$ | — | — | — | — | [531] | — | — | — | — | — | — |
| $CF_2{=}CHCl$ | — | — | — | — | [530] | — | — | — | — | — | — |
| $CF_2{=}CFCl$ | [520] | [520] | [532] | — | [527] [532] | — | [520] | — | — | — | — |
| $CF_2{=}CF_2$ | [520] | [520] | — | — | [521] | — | [520] | — | [521] | — | — |
| $CH_2{=}CH{-}CH_3$ | — | — | — | — | — | [536] | — | [537] | — | — | — |
| $CHCl{=}CH{-}CH_3$ | — | — | [519] | — | — | — | — | — | — | — | — |
| $CH_2{=}C{=}CH_2$ | — | — | — | — | [526] | — | — | — | — | — | — |
| $CH_2{=}CH{-}CH_2Cl$ | — | — | — | — | [526] | — | — | — | — | — | [534] |
| $CF_2{=}CH{-}CH_3$ | — | — | — | — | [523] | — | — | — | — | — | — |
| $CH_2{=}CH{-}CF_3$ | — | — | [533] | [518] | [533] | — | — | — | — | — | — |
| $CF_2{=}CH{-}CF_3$ | — | — | — | — | [529] | — | — | — | — | [529] | — |
| $CF_2{=}CF{-}CF_3$ | — | — | — | — | [525[ | — | — | — | — | — | — |
| $CH_2{=}C(CH_3)_2$ | — | — | — | — | — | — | — | [537] | — | — | — |

[a] Boldface halogen splits off during the dissociation.
[b] Boldface carbon combines with the paraffin radical carbon.

**Table 48. Addition of Haloparafins to Acetylenes**

| | Haloparaffin | | | |
|---|---|---|---|---|
| Acetylene[a] | $CCl_3I$ | $CF_3I$ | $C_2F_5I$ | $C_3F_7I$ |
| $CH{\equiv}CH$ | — | **[540]** | **[543,544]** | **[539]** |
| $CH{\equiv}C{-}CH_3$ | — | **[545,546]** | — | — |
| $CH{\equiv}C{-}CF_3$ | **[541]** | **[542,546]** | — | **[546]** |

*a* Boldface carbon combines with the paraffin radical carbon.

(320)

$$CF_2{=}CFI \xrightarrow{CH_2=CH_2} CF_2{=}CFCH_2CH_2I \quad 67\%$$
$$CF_2{=}CFI \xrightarrow{CH_2=CHF} CF_2{=}CFCH_2CHFI \quad 50\%$$
$$CF_2{=}CFI \xrightarrow{CH_2=CF_2} CF_2{=}CFCH_2CF_2I \quad 24\%$$
$$CF_2{=}CFI \xrightarrow{h\nu} CF_2{=}CFCF_2CFI_2 \quad 50\%$$
$$CF_2{=}CFI \xrightarrow{CF_2=CCl_2} CF_2{=}CFCF_2CCl_2I \quad 5\%$$
$$CF_2{=}CFI \xrightarrow{CF_2=CHCl} CF_2{=}CFCHClCF_2I \quad 4\%$$
$$CF_2{=}CFI \xrightarrow{CF_2=CHF} CF_2{=}CFCHFCF_2I \quad 39\%$$
$$CF_2{=}CFI \xrightarrow{170\text{–}180°C} \begin{array}{l} CF_2{-}CF \\ | \quad\ \ \| \\ CF_2{-}CF \end{array}$$

**[517]**

Halogenoparaffins and fluorohaloparaffins also add to *acetylene* and its homologs to form fluorinated olefins, which in turn may be either oxidized to acids or ketones, or subjected to many addition reactions. Consequently, the addition of halo- and fluorohaloparaffins to acetylenes is of much synthetic use (Table 48).

Another type of linear addition is the addition of alcohols, ethers, and aldehydes to fluorinated olefins in the way that the free radical generated by irradiation or organic peroxides from the addends joins one carbon and a hydrogen atom the other carbon of the double bond. Alcohols give homologous alcohols, ethers give different ethers, and aldehydes give ketones **[547,548]**.

(321)

$$\begin{array}{l} CF_2{-}CCl \\ | \qquad \| \\ CF_2{-}CCl \end{array} \xrightarrow[\gamma\ \text{rays}]{C_2H_5OH} \begin{array}{l} CF_2{-}C(Cl)(H) \\ | \qquad\ \ | \\ CF_2{-}C(Cl){-}CHCH_3 \\ \qquad\qquad\quad | \\ \qquad\qquad\ \ OH \end{array} \ 18\% + \begin{array}{l} CF_2{-}C(Cl)(H) \\ | \qquad\ \ | \\ CF_2{-}C(Cl)(H) \\ \qquad\ \ | \\ \quad HOCHCH_3 \end{array} \ 36\% + \begin{array}{l} CF_2{-}CCl \\ | \qquad \| \\ CF_2{-}C{-}CHCH_3 \\ \qquad\qquad\ \ | \\ \qquad\qquad OH \end{array} \ 18\%$$

**[547]**

(322)

$$\begin{array}{l} CF_2{-}CF \\ | \qquad \| \\ CF_2{-}CF \end{array} \xrightarrow[\gamma\ \text{rays}]{CH_3CHO} \begin{array}{l} CF_2{-}C(F)(H) \\ | \qquad\ \ | \\ CF_2{-}C(F){-}COCH_3 \end{array} \ 20\% + \begin{array}{l} CF_2{-}C(F)(H) \\ | \qquad\ \ | \\ CF_2{-}C(F)(COCH_3) \end{array} \ 5\%$$

**[548]**

*Cycloadditions*

Cycloadditions are reactions in which ring compounds are formed. Three-, four-, five-, six-, and, rarely, eight-membered rings result, depending on the reaction components.

*Formation of Three-Membered Rings.* A three-center mechanism is proposed for the reaction of carbenes with olefins or fluoro-olefins. The best generator of carbene (or methylene) is diazomethane. Fluorochlorocarbene, difluorocarbene, and higher carbenes are produced by different types of decomposition **[549–551]**.

(323) Cyclohexene → [$(CH_3)_3SnCF_3$, NaI, $(CH_2OCH_3)_2$; 80°C, 12 hr; [550]] → difluoronorcarane ($CF_2$) 73%

Cyclohexene → [$CCl_2FCO_2CH_3$, MeOH, $Et_2O$; reflux 0.5 hr; [549]] → 35% fluorochloronorcarane ($CFCl$)

Cyclohexene → [$CCl_2FCOCCl_2F$, MeONa; 35–40°C] → 38% fluorochloronorcarane ($CFCl$)

$$(CH_3)_2C{=}CH_2 + CF_2\langle N{=}N \rangle \xrightarrow[20°C,\ 20\ hr]{h\nu} (CH_3)_2C\langle CH_2{-}CF_2 \rangle \quad 71\% \qquad [551] \tag{324}$$

*Formation of Four-Membered Rings.* Cycloadditions leading to four-membered rings are especially frequent with fluoro-olefins, and occur in preference to other reactions. The reaction can take place between two different fluoro-olefins, or between two molecules of the same fluoro-olefin (cyclodimerization). It is not entirely clear whether the reaction takes place by a four-center simultaneous cyclization, or by two consecutive reactions, formation of a biradical, and its subsequent ring-closure. The reaction always takes place by head-to-head combination, is accelerated by heat, and does not seem to be affected by light and peroxides like the linear chain reactions described previously.

In the reaction with dienes having a transoid system of double bonds, four-membered rings are formed exclusively, whereas with cisoid olefins, the *Diels–Alder* reaction prevails **[552]**.

(325) [552]

$CF_2{=}CCl_2$ + $CH_2{=}CH{-}CH{=}CH_2$ (80°C) → $CF_2{-}\dot{C}Cl_2$ / $CH_2{-}\dot{C}H{-}CH{=}CH_2$ (biradical) → $CF_2{-}CCl_2$ / $CH_2{-}CH{-}CH{=}CH_2$ (cyclobutane) 92%

$CF_2{=}CCl_2$ + cyclopentadiene (80°C) → bicyclic cyclobutane adduct ($CF_2$, $CCl_2$) 9% + bicyclo[2.2.1]heptene adduct ($CF_2$, $CCl_2$) 44%

In the dimerization of nonsymmetrical fluoro-olefins, both *cis* and *trans* derivatives of cyclobutane are formed **[553]**. An interesting dimer of perfluorobutadiene once formulated as a system of three four-membered rings has recently been recognized as a compound containing five-membered rings instead **[554]**.

(326) 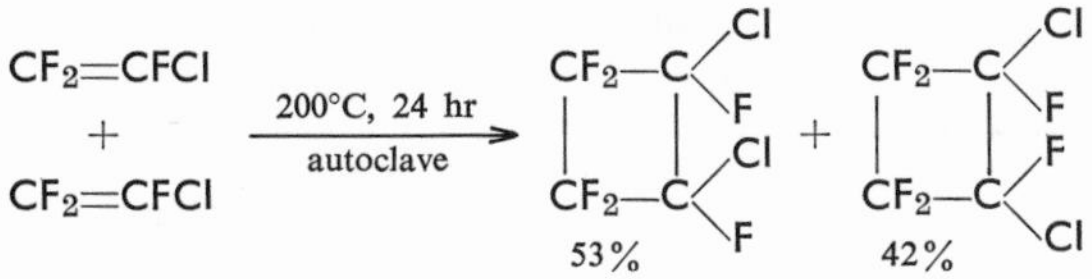
CF2=CFCl
+
CF2=CFCl
200°C, 24 hr
autoclave
53%
42%
**[553]**

(327) 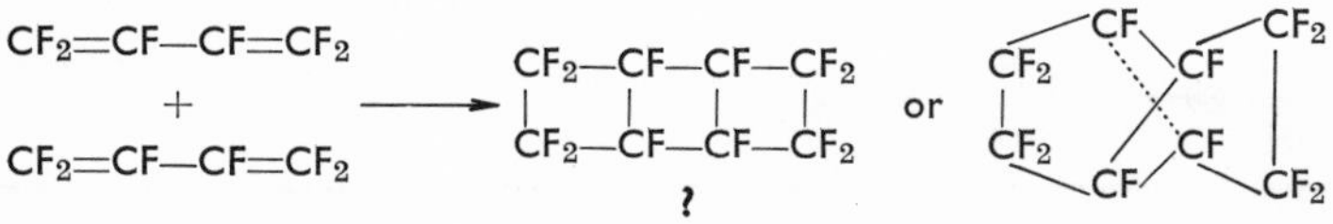
CF2=CF—CF=CF2
+
CF2=CF—CF=CF2
or
?
**[554]**

Four-membered rings are also formed in additions of fluoro-olefins to acetylenes. In acetylenic olefins, both the triple bond and the double bond react at approximately same rates **[555,556]**.

(328) 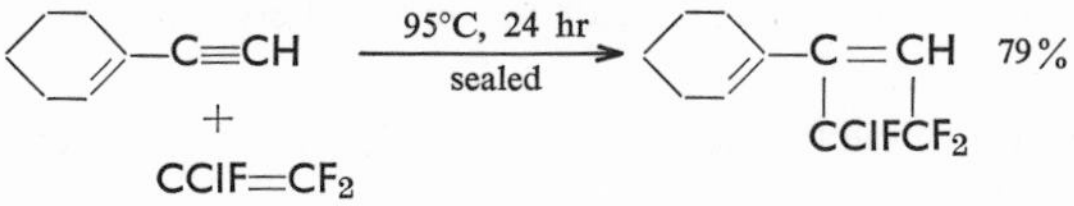
95°C, 24 hr
sealed
79%
CClF=CF2
**[555]**

(329) 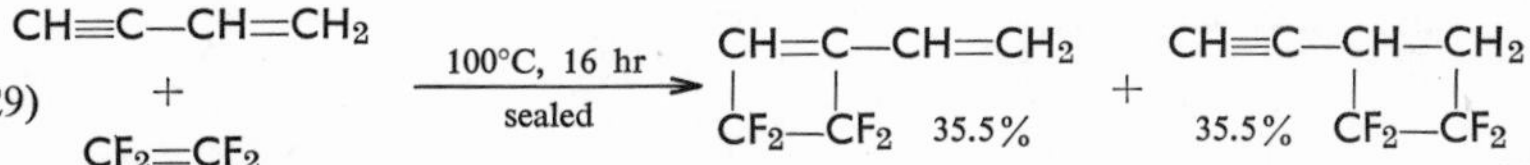
CH≡C—CH=CH2
+
CF2=CF2
100°C, 16 hr
sealed
35.5%
35.5%
**[556]**

Cycloadditions leading to four-membered rings also result from combinations of fluoro-olefins with compounds containing double bonds between two hetero-atoms. An example is the reaction of fluoro-olefins with nitroso compounds, in which linear head-to-tail polymerization competes with the cycloaddition **[557]**.

(330) 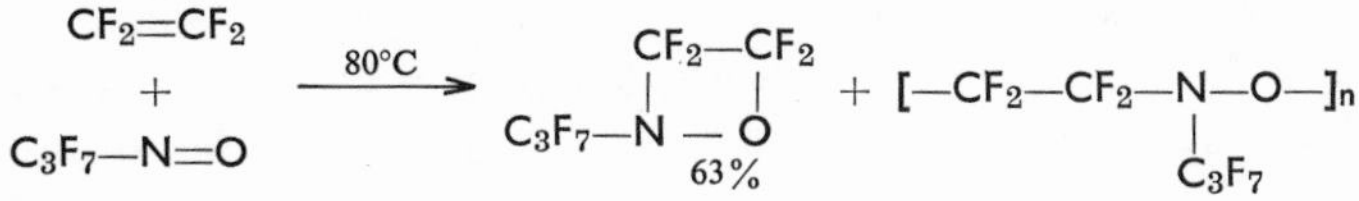
CF2=CF2
+
C3F7—N=O
80°C
63%
**[557]**

*Formation of Five-Membered Rings*. Five-membered rings are formed relatively rarely. One example is the addition of diazomethane to 3,3,3-trifluoropropene, which gives a pyrazoline derivative. This decomposes to a cyclopropane compound **[558]**.

(331) 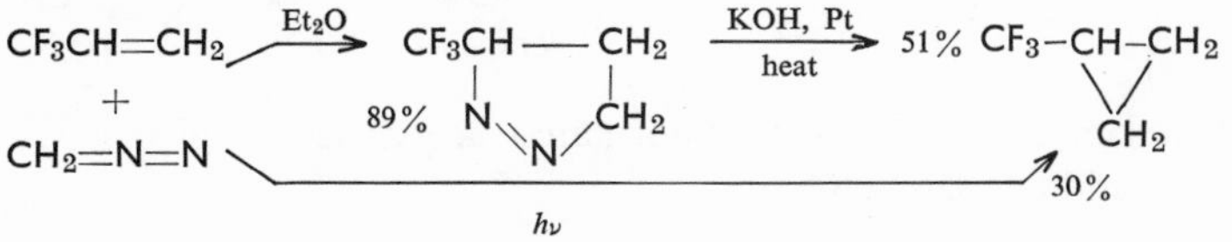
CF3CH=CH2
+
CH2=N=N
Et2O
89%
KOH, Pt
heat
51%
30%
hν
**[558]**

**Table 49. Results of the Diels–Alder Reaction of 1,1-Difluorotetrachloro-2,4-cyclopentadiene with Some Unsaturated Compounds [559]**

| Unsaturated compound | Yield of the adduct, % | Unsaturated compound | Yield of the adduct, % |
|---|---|---|---|
| $C_6H_5$—CH=CH$_2$ | 95 | CH$_2$=CH·COOH | 72 |
| Cl—$C_6H_4$—CH=CH$_2$ | 75 | CH—COOH ‖ CH—COOH | 88 |
| CH=CH—CH$_2$—CH=CH (cyclopentadiene) | 74 | O=$C_6H_4$=O | 56 |

*Formation of Six-Membered Rings* (*Diels–Alder Reaction*). The abundance of six-membered ring products resulting from the *Diels–Alder* reaction of fluorinated olefins and dienes is decreased by the competitive reaction of fluoro-olefins, by cycloaddition and cyclodimerization. Nevertheless, quite a few examples of regular four-center diene synthesis can be quoted (Table 49) **[559,560]**.

In the reaction of cyclic dienes, both endo and exo products are obtained **[561]**:

(332) cyclopentadiene + CF—CF$_2$ ‖ | CF—CF$_2$ —(90°C, 24 hr, sealed)→ endo adduct (F, F$_2$, F, F$_2$) 30% + exo adduct (F, F$_2$, F, F$_2$) 30% **[561]**

Frequently, acetylenic compounds are used as dienophiles. Some of the products obtained in this way readily aromatize on heating **[559,562]**.

(333) Cl$_4$C$_5$F$_2$ + CO$_2$CH$_3$—C≡C—CO$_2$CH$_3$ ⟶ bicyclic adduct (Cl, Cl, Cl, Cl, CF$_2$, CO$_2$CH$_3$, CO$_2$CH$_3$) 46% —(480°C)→ tetrachlorodimethyl phthalate (Cl, Cl, Cl, Cl, CO$_2$CH$_3$, CO$_2$CH$_3$) 83% **[561]**

(334) F, F, F, F$_2$, F$_2$, F (hexafluorocyclohexadiene) + CF$_3$—C≡C—CF$_3$ —(225°C, 18 hr, sealed)→ bicyclic adduct (F, F, F, F, CF$_2$, CF$_2$, CF$_3$, CF$_3$) 67% —(500°C)→ F, F, F, F, CF$_3$, CF$_3$ (aromatic) 97% **[562]**

Even benzene reacts with fluorinated acetylenes to give a series of interesting adducts **[563,564]**:

(335) 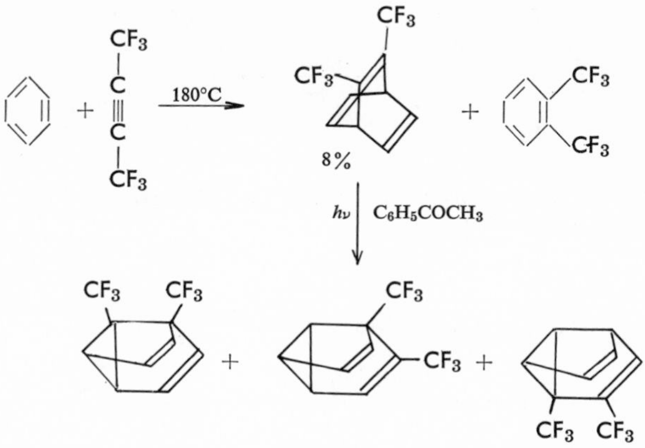 [563, 564]

Tetrafluorobenzyne acts as a dienophile in the reaction with thiophene and N-methylpyrrol **[565]**:

(336) 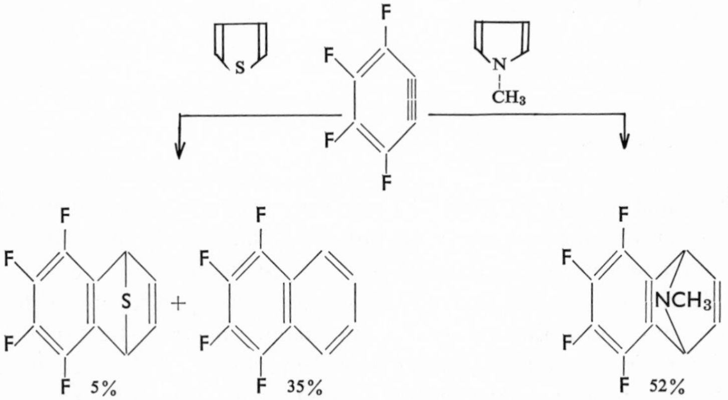 [565]

Also, heterocyclic compounds are accessible by the *Diels–Alder* reaction of fluorinated dienes and nitriles or nitroso compounds, respectively. The former reaction gives derivatives of pyridine **[566],** the latter those of dihydro-oxazine **[567]**:

(337) 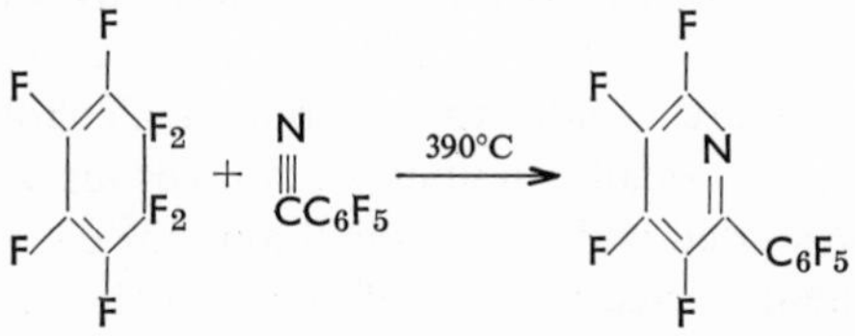 [566]

$$CF_2{=}CF{-}CF{=}CF_2 + O{=}NCF_3 \xrightarrow[-180° \to 70°C]{\text{exothermic}} \text{cyclo-}(CF_2{-}CF{=}CF{-}CF_2{-}N(CF_3){-}O{-}) \quad 57.5\% \tag{338}$$

[567]

Cycloadditions are involved in several interesting reactions of hexafluoro-2-butyne **[568–571]**:

(339)

$CF_3C{\equiv}CCF_3$:
- $I_2$, 260°C, 1000 atm, 15 hr **[569]** → hexakis(trifluoromethyl)benzene, 71%
- 320°C, 31 hr **[570]** → octakis(trifluoromethyl)cyclooctatetraene, 37%
- P(red), 200°C, 8 hr, sealed **[568]** → hexakis(trifluoromethyl)-1,4-diphosphabarrelene, 43%
- $Fe(CO)_5$, 110°C, 16 atm, sealed **[571]** → tricarbonyl[tetrakis(trifluoromethyl)cyclopentadienone]iron complex, 60%

## ELIMINATIONS

Eliminations are very useful in fluorine chemistry, since they are the most general method for the preparation of fluorinated olefins and acetylenes. Not unfrequently, eliminations are used for recovering olefins from their dibromides—a common way of isolation and purification of unsaturated compounds. Eliminations occur mainly as *trans*-1,2-eliminations. 1,1-Eliminations leading to carbenes are exceptional, and 1,3-eliminations giving cyclopropane derivatives are rare. Quite a few examples are known of intermolecular eliminations resulting in coupling of fluorinated halides.

### Dehalogenations

Although vicinal halogen atoms can be split out from a molecule of a fluorinated compound by heating to a high enough temperature or by hydrogen and a metal catalyst, by far the most common reagents for accomplishing the conversion of vicinal dihalo derivatives to olefins are zinc and magnesium.

Dehalogenations are usually carried out in alcohols or ethers using zinc dust activated by small amounts of anhydrous zinc chloride. The reaction should be started with just a small amount of the dihalide and by local heating of the unstirred mixture of zinc and the organic solution. As soon as the reaction sets off, heating should be stopped, stirring started, and the dihalide added gradually at such a rate as to keep the exothermic reaction vigorous. Vicinal dihaloparaffins give olefins **[572]**, vicinal dihalo-olefins give acetylenes **[573]**.

(340) $$CFCl_2CF_2Cl \xrightarrow[70°C]{Zn,\ EtOH} CFCl{=}CF_2 \quad 92\%$$ [572]

(341) $$CF_3CCl{=}CClCF_3 \xrightarrow[40°C,\ reflux]{Zn,\ EtOH} CF_3C{\equiv}CCF_3\ 55\% + CF_3CCl{=}CHCF_3\ (cis,\ trans)\ 30\% + CF_3CH_2CH_2CF_3\ 1.6\%$$ [573]

Fluorine atoms resist dehalogenation. There are, nonetheless, examples of 1,2-dehalogenation of vicinal fluoroiodides, fluorobromides, and even fluorochlorides. These eliminations compete successfully with the formation of Grignard reagents from vicinal fluorohalides and magnesium **[574]**.

(342) $$CF_3CH_2I \xrightarrow{Mg,\ Et_2O} CF_2{=}CH_2 \quad 90\%$$ [574]

Vicinal fluorine atoms are eliminated only at very high temperatures using iron or nickel. This reaction is extremely important in perfluorohydroaromatic compounds, which are thus converted to perfluoroaromatics. To a certain extent, this *defluorination* resembles the dehydrogenation of hydroaromatic compounds to give aromatic derivatives **[575]**.

(343) Perfluoro(methylcyclohexane) (ring CF(CF₃), CF₂, CF₂, CF₂, CF₂, CF₂) $\xrightarrow{Fe,\ 500°C}$ perfluorotoluene (ring C(CF₃), CF, CF, CF, CF, CF) 25% [575]

Intermolecular dehalogenations are used for the coupling of fluoroalkyl or fluoroalkenyl iodides in the Wurtz–Fittig reaction **[576]**. Even fluorinated aromatic halides can be coupled together, or to fluoroaliphatic halides using copper (p. 129).

(344) $$CF_2{=}CFCF_2I \xrightarrow[reflux]{Zn,\ dioxane} CF_2{=}CFCF_2CF_2CF{=}CF_2 \quad 83\%$$ [576]

## Dehydrohalogenations

Dehydrohalogenations are used less frequently as a method for the preparation of fluorinated olefins. One of the reasons for this is that they are usually not as easy to carry out as dehalogenations. There are, however, several examples of preparatively important dehydrohalogenations.

Thermal 1,1-elimination of hydrogen chloride from chlorodifluoromethane gives difluorocarbene, which dimerizes to tetrafluoroethylene **[577]**:

(345) $$CHClF_2 \xrightarrow[0.5\ atm]{650\text{–}800°C} [CF_2] \longrightarrow CF_2{=}CF_2 \quad \text{conv. 25–30\%, yield 90\%}$$ **[577]**

Much more common are 1,2-eliminations carried out by using alkalies **[578]** or organic bases **[580]**. The latter reaction is one of the routes to perfluorodimethylketene **[579,581]**.

(346) $$C_6H_5CFClCHF_2 \xrightarrow[230\text{–}250°C]{NaOH} C_6H_5CF{=}CF_2 \quad 17\%$$ **[578]**

(347)
$$(CF_3)_2CHCOCl \xrightarrow[-78°C]{Et_3N} (CF_3)_2C{=}C{=}O \quad 24\% \quad [580]$$
$$(CF_3)_2CHCOBr \xrightarrow[-78°C]{Et_3N} (CF_3)_2C{=}C{=}O \quad 57\% \quad [580]$$
$$(CF_3)_2CHCOF \xrightarrow[100°C,\ 16\ hr]{KF} (CF_3)_2C{=}C{=}O \quad 35\% \quad [579]$$
$$(CF_3)_2CHCO_2H \xrightarrow[reflux\ 4\ hr]{P_2O_5} (CF_3)_2C{=}C{=}O \quad 94\% \quad [581]$$

Occasionally, even hydrogen fluoride can be split out, sometimes preferentially to hydrogen chloride **[582]**. Elimination of hydrogen fluoride from fluorinated cyclohexanes gives perfluoroaromatics **[583]**.

(348)
$$CF_2ClCH_3 \xrightarrow[NiO,\ 400°C]{-HCl} CF_2{=}CH_2 \quad 99.4\%$$
$$CF_2ClCH_3 \xrightarrow[Al_2O_3,\ 300°C]{-HF} CFCl{=}CH_2 \quad 91\%$$
**[582]**

(349) $$C_6F_8H_4 \text{ (tetrahydro-F-cyclohexane)} \xrightarrow[reflux\ 4.5\ hr]{50\%\ KOH} C_6F_5H \quad 52.8\%$$ **[583]**

## Decarboxylations

A unique reaction is the elimination of carbon dioxide from alkaline salts of perfluoroalkanecarboxylic acids. According to the conditions used, either 1-hydrylperfluoroalkanes can be obtained, when the reaction is carried out in a glycol, or perfluoro-olefins can result from dry distillation. Fluoro-olefins with terminal double bonds are the primary products. Fluoro-olefins with internal double bonds also isolated from the reaction mixtures are products of rearrangement of the double bond due to the alkaline fluoride catalysis **[584,585]**.

(350)
$$CF_3CF_2CF_2CHF_2\ 84\% \xleftarrow[170\text{–}190°C]{(CH_2OH)_2}_{[584]} CF_3CF_2CF_2CF_2CO_2K \xrightarrow[{[585]}]{165\text{–}200°C} 7\%\ CF_3CF_2CF{=}CF_2 + 29\%\ CF_3CF{=}CFCF_3$$

The reaction is of great preparative importance, since perfluorocarboxylic acids are readily available by the electrochemical fluorination process.

### Dehydration

Dehydration of polyfluoro and perfluoro hydroxy compounds is very difficult, and heating with concentrated sulfuric acid and even phosphorus pentoxide fails in some instances. An example of practical application is the dehydration of hexafluoroisobutyric acid, which gives perfluorodimethyl ketene **[581]** [Equation (347)].

## MOLECULAR REARRANGEMENTS

Molecular rearrangements are quite frequent in fluorine chemistry, and some of them have been mentioned already (p. 88). The reactions comprise double-bond shifts, rearrangements of halogen atoms, rearrangements of carbon skeleton, shifts from oxygen and nitrogen to carbon, etc. Only those rearrangements that have some synthetic use will be discussed. A *double-bond shift* catalyzed by fluoride ions occurs in the decarboxylation of sodium or potassium salts of perfluoroalkanecarboxylic acids and leads to perfluoroolefins with internal double bonds [Equation (350)].

One of rearrangements peculiar to fluorine compounds is a *shift of fluorine*. Perfluorodimethylketene is thus converted to perfluoromethacryloyl fluoride **[586]**. Of much greater preparative importance are shifts of fluorine to form geminal or trigeminal polyfluoro compounds. These reactions are catalyzed by aluminum chloride or bromide, and provide a way to compounds which are difficult to prepare otherwise **[587,588]**.

(351)
$$(CF_3)_2C{=}C{=}O \xrightarrow[300°C]{NaF} CF_2{=}C(CF_3){-}COF \quad \text{conv. } 34\%, \text{ yield } 92\% \quad [586]$$

(352)
$$CF_2BrCHClF \xrightarrow[50°C]{AlCl_3} CF_3CHClBr\ 90\% \quad [587]$$
$$CF_2BrCClBrF \xrightarrow[90°C]{AlBr_3} CF_3CClBr_2\ 90\% \quad [588]$$
$$CF_3CClBr_2 \xrightarrow{Na_2SO_3,\ NaOH} CF_3CHClBr$$

*Isomerizations of carbon skeletons* occur frequently. Migration of perfluoroalkyl groups attached to fluorinated pyridine is catalyzed by fluoride ions [589], and the isomerization of hexafluorobenzene to hexafluorobicyclo(0,2,2)hexadiene-2,5 [590, 591] and of hexakis(trifluoromethyl)benzene to three nonaromatic isomers [592] are achieved by irradiation.

(353) $(CF_3)_2CF$, $CF(CF_3)_2$, F, F, $CF(CF_3)_2$ pyridine $\xrightarrow[130°C]{KF}$ F, $CF(CF_3)_2$, F, $(CF_3)_2CF$, $CF(CF_3)_2$ pyridine [589]

(354) $C_6F_6$ $\xrightarrow{h\nu\ 2537 Å,\ 20°C}$ hexafluorobicyclo(0,2,2)hexadiene 60% [590]

(355) $C_6(CF_3)_6$ $\xrightarrow[2540\ Å]{h\nu}$ hexakis(trifluoromethyl)bicyclohexadiene 54% + hexakis(trifluoromethyl)benzvalene 43% + hexakis(trifluoromethyl)prismane 3% [592]

Treatment of fluorinated acyl chlorides with diazomethane and the subsequent rearrangement of the diazo ketones by silver salts in appropriate solvents (*Wolff rearrangement and Arndt–Eistert reaction*) provides for the extension of the carbon chain by one carbon atom [593]. The method is not general and some fluoroacyl chlorides give irregular results [594].

(356) $CF_3COCl \xrightarrow[0-20°C]{CH_2N_2} CF_3COCHN_2$ (62.5%) $\xrightarrow{Ag_2O,\ C_2H_5OH} CF_3CH_2CO_2C_2H_5$ (40.4%) [593]

In the mechanistically related *Curtius rearrangement*, fluorinated acyl chlorides are converted to fluoroacyl azides, which, depending on the reaction conditions, can be transformed to fluorinated isocyanates, urethanes, or amides [595,596].

(357)

$CF_3CF_2CF_2COCl \xrightarrow[20°C]{NaN_3} CF_3CF_2CF_2CON_3$

[596] $CF_3CF_2CF_2COCl \xrightarrow[55-65°C]{NaN_3}$ $CF_3CF_2CONH_2$ (10%) + 75%

$CF_3CF_2CF_2CON_3 \xrightarrow[110°C]{C_7H_8} CF_3CF_2CF_2NCO$ (82%) $\xrightarrow{C_2H_5OH} CF_3CF_2CF_2NHCO_2C_2H_5$ (91%) [595]

*Hofmann degradation*, leading regularly from amides to amines, gives with fluorinated compounds products quite different from those expected **[597,598]**. Depending on the reagents and reaction conditions, perfluoroalkyl halides, or monohydryl perfluoroalkanes, or perfluoroalkenes are obtained. Under special conditions, even perfluoroalkylisocyanate was isolated.

(358)

$$CF_3CF_2CF_2CONH_2 \xrightarrow{\text{NaOI, distn.}} CF_3CF_2CHF_2 \quad [597] \quad 25\text{–}40\%$$

$$CF_3CF_2CF_2CONH_2 \xrightarrow{\text{NaOCl, 105°C}} CF_3CF_2CF_2Cl + C_3F_6 + C_3HF_7 + C_2HF_5 \quad [597]$$

$$CF_3CF_2CF_2CONH_2 \xrightarrow{\text{NaOBr, 95°C}} CF_3CF_2CF_2Br \quad 65\text{–}70\%$$

$$CF_3CF_2CF_2CONH_2 \xrightarrow{\text{Ag}_2\text{O, Et}_2\text{O, reflux}} CF_3CF_2CF_2CONHAg \; 98\% \quad [598]$$

$$CF_3CF_2CF_2CONHAg \xrightarrow{Br_2} CF_3CF_2CF_2CONHBr \; 75\%$$

$$CF_3CF_2CF_2CONHBr \xrightarrow[\text{5–10°C}]{\text{NaOH}} [CF_3CF_2CF_2\text{—}CO(\text{—}N\text{—}Br)]^{\ominus} Na^{\oplus} \; 99\%$$

$$[CF_3CF_2CF_2\text{—}CO(\text{—}N\text{—}Br)]^{\ominus} Na^{\oplus} \xrightarrow{\text{H}_2\text{O, 100°C, 0.5 hr}} CF_3CF_2CF_2Br \; 91\%$$

$$[CF_3CF_2CF_2\text{—}CO(\text{—}N\text{—}Br)]^{\ominus} Na^{\oplus} \xrightarrow{\text{165–170°C}} CF_3CF_2CF_2NCO \; 83\%$$

## PYROREACTIONS

Pyroreactions are important, especially in the chemistry of tetrafluoroethylene. The polymer depolymerizes on heating to give monomeric tetrafluoroethylene, which dimerizes to acyclic dimer (perfluorocyclobutane), and, in addition, gives perfluoroisobutylene and perfluoropropylene **[599–601]**.

(359)

$$(CF_2CF_2)_n \xrightarrow[\text{350 mm}]{\text{600–700°C}} CF_2{=}CF_2 \; 32.5\% \xrightarrow{[599]} CF_2{=}CFCF_3 + (CF_3)_2C{=}CF_2 \; 47.5\% + \begin{matrix} CF_2\text{—}CF_2 \\ | \quad\quad | \\ CF_2\text{—}CF_2 \end{matrix}$$

$$\begin{matrix} CF_2\text{—}CF_2 \\ | \quad\quad | \\ CF_2\text{—}CF_2 \end{matrix} \xrightarrow{\text{700–725°C}} CF_2{=}CFCF_3 \; 20\% + (CF_3)_2C{=}CF_2 \; 70\% \quad [600]$$

$$CF_2{=}CFCF_3 \xrightarrow{\text{Ni tube, 750°C}} (CF_3)_2C{=}CF_2 \quad [601]$$

In the field of fluorinated heterocyclics, pyrolysis occurs with perfluorooxazetidines, which are converted to carbonyl fluoride and perfluoroalkyliminomethylene **[602]**.

(360)

$$\begin{matrix} C_3F_7N\text{—}CF_2 \\ | \quad\quad | \\ O\text{—}CF_2 \end{matrix} \xrightarrow{\text{550–600°C}} \begin{matrix} 96\% \; C_3F_7N{=}CF_2 \\ + \\ 102\% \; O{=}CF_2 \end{matrix}$$

$$\text{polymer } \left[\begin{matrix} | \quad\quad | \\ C_3F_7N \quad CF_2 \\ | \quad\quad | \\ O\text{—}CF_2 \end{matrix}\right] \xrightarrow[10^{-3}\text{ mm}]{\text{450–500°C}} \begin{matrix} C_3F_7N{=}CF_2 \; 60\% \\ + \\ O{=}CF_2 \; 60\% \end{matrix} \quad [602]$$

**Table 50. Survey of Interconversions of Main Types of Organic Fluorine Compounds**

| From fluorinated: | To fluorinated: | | | | |
|---|---|---|---|---|---|
| | Paraffins, cycloparaffins | Olefins, cycloolefins | Acetylenes | Aromatics | Halogen-derivatives[a] |
| Paraffins, cycloparaffins | $F_2$; rearr. | — | — | Fe or Ni, heat[b] | $X_2$ |
| Olefins, cycloolefins | $H_2$ catal. | Rearr.; RMgX, RLi | Zn | Fe or Ni, heat[b] | $X_2$; HX; RX; $AlX_3$ |
| Acetylenes | $H_2$ catal. | $H_2$ catal. | — | Trimerization | $X_2$; HX; RX |
| Aromatics | $H_2$ catal.[b] | — | — | Rearr.; RMgX, RLi | $X_2$ |
| Halogen derivatives | $H_2$ catal.; Na; Zn; Fe; $Na_2SO_3$ | Zn; Mg; KOH | Zn; KOH | — | $F_2$; $X_2$; Rearr. $AlX_3$ |
| Nitro compounds | — | — | — | — | — |
| Alcohols, phenols | — | $H_2SO_4$; $P_2O_5$ | — | — | HX; $PX_3$ |
| Aldehydes | — | Wittig | — | — | — |
| Ketones | — | Wittig | — | — | $PX_5$ |
| Acids | Na salts, heat | Na salts, heat | — | — | Ag salts, $X_2$ |
| Acyl halides | — | — | — | — | — |
| Amides | — | — | — | — | — |
| Nitriles | — | — | — | — | — |
| Esters | — | — | — | — | — |
| Sulfoesters | — | — | — | — | KX |
| Amines | — | — | — | — | $NaNO_2$+HX |

[a] X=Cl, Br, I.
[b] Only six-membered rings.

**Table 50 (Continued)**

| From fluorinated: | To fluorinated: | | | | |
|---|---|---|---|---|---|
| | Nitro compounds | Alcohols, phenols | Aldehydes | Ketones | Acids |
| Paraffins, cycloparaffins | $N_2O_4$ | — | — | $H_2SO_4$[c] | $O_2$; $H_2SO_4$ |
| Olefins, cycloolefins | $N_2O_4$ | — | $O_3$ | $CrO_3$; $KMnO_4$; $H_2SO_4$ | $KMnO_4$ |
| Acetylenes | — | — | $H_2O$ | $H_2O$; $CrO_3$ | $KMnO_4$ |
| Aromatics | $HNO_3$ | KOH | $X_2$, hydrol.; $CrO_2Cl_2$ | — | $HNO_3$ dil.; $CrO_3$; $KMnO_4$ |
| Halogen derivatives | $AgNO_2$ | KOH | $Pb(NO_3)_2$ | — | $O_2$; $H_2SO_4$; Mg+$CO_2$ |
| Nitro compounds | $HNO_3$ | — | KOH; $H_2SO_4$ | — | $H_2SO_4$ |
| Alcohols, phenols | — | — | $MnO_2$; $Cl_2$ | — | $CrO_3$ |
| Aldehydes | — | $LiAlH_4$ | — | — | — |
| Ketones | — | $H_2$ catal.; ROH $LiAlH_4$; $NaBH_4$ Zn; RMgX | — | Aldol condens. | — |
| Acids | — | $LiAlH_4$; RMgX | $LiAlH_4$ | RMgX; RLi | — |
| Acyl halides | — | $LiAlH_4$ | $H_2$/Pd | $R_2Cd$; RMgX | $H_2O$ |
| Amides | — | $H_2$ catal. | $LiAlH_4$[c] | — | $H_2O$ |
| Nitriles | — | — | $LiAlH_4$ | RMgX | $H_2O$ |
| Esters | — | $LiAlH_4$; RMgX | $LiAlH_4$[c] | RMgX | $H_2O$ |
| Sulfoesters | — | $H_2O$ | — | — | — |
| Amines | Peracids | — | — | — | — |

[c] Only in special cases.

**Table 50 (Continued)**

| From fluorinated: | To fluorinated: Acyl halides | Amides | Nitriles | Esters | Sulfoesters | Amines |
|---|---|---|---|---|---|---|
| Paraffins, cycloparaffins | — | — | — | — | — | — |
| Olefins, cycloolefins | $O_2$ | — | $NH_3$ | $RCO_2H$ | — | $RNH_2$; $R_2NH$ |
| Acetylenes | — | — | — | $RCO_2H$ | — | — |
| Aromatics | — | — | — | — | — | $NH_3$; $RNH_2$; $R_2NH$ |
| Halogen derivatives | $O_2$; oleum | — | KCN | — | — | $NH_3$; $RNH_2$; $NHR_2$ |
| Nitro compounds | — | — | — | — | — | $H_2$ catal.; Sn |
| Alcohols, phenols | — | — | — | $RCO_2H$; RCOX | $RSO_2Cl$ | — |
| Aldehydes | — | — | — | — | — | — |
| Ketones | — | — | — | — | — | — |
| Acids | $PX_3$; $POX_3$; $PX_5$; $SOCl_2$ | — | — | ROH; olefins | — | Curtius, Schmidt |
| Acyl halides | — | $NH_3$ | — | ROH | — | Curtius |
| Amides | — | — | $P_2O_5$ | ROH | — | $LiAlH_4$; $B_2H_6$; Hofmann |
| Nitriles | — | $H_2O$ | — | ROH | — | $H_2$ catal. $LiAlH_4$, $NaBH_4$ |
| Esters | — | $NH_3$ | — | Claisen condens. | — | — |
| Sulfoesters | — | — | — | — | — | — |
| Amines | — | $H_2O$[c] | — | — | — | — |

[c] Only in special cases.

Practical applications of reactions of fluorinated compounds for synthetic or preparative purposes, i.e., for converting one type of compound to another, have been summarized in Table 50.

*Chapter 9*

# Fluorinated Compounds as Chemical Reagents

Several fluorinated compounds react only transiently with organic substrates and do not participate in forming the final products. One of the most useful reagents of this kind is *trifluoroacetic anhydride*. It is commonly used for esterifications of alcohols that are too acidic or of sterically hindered acids. It converts organic acids to mixed anhydrides which are readily cleaved to an acyl carbonium and a trifluoroacetate ion **[603,604]**.

(361) $$2,6\text{-}(CH_3)_2C_6H_3\text{—}CO_2H \xrightarrow[20°C,\ 2\ hr]{CH_3OH,\ (CF_3CO)_2O} 2,6\text{-}(CH_3)_2C_6H_3\text{—}CO_2CH_3 \quad 93\%$$ **[603]**

Another fluorinated reagent of much practical value is *trifluoroperoxy-acetic acid*, which is usually prepared *in situ* by mixing 90% hydrogen peroxide with trifluoroacetic anhydride. This reagent is very selective, and is especially suited for converting primary amines to nitro compounds **[605]**, and olefins to epoxides **[606]** and *trans*-diols (after hydrolysis of the unstable intermediates, vicinal hydroxytrifluoroacetates) **[607]**, and ideal for converting ketones to esters in the *Baeyer–Villiger* reaction. The latter reaction is important not only from the synthetic point of view **[608,609]**, but also for theoretical studies of migratory aptitudes of various groups **[610]** (Table 51).

(362) $$C_6F_5NH_2 \xrightarrow[reflux\ 18\ hr]{(CF_3CO)_2O,\ 90\%\ H_2O_2,\ CH_2Cl_2} C_6F_5NO_2 \quad 85\%$$ **[605]**

(363) $$\text{cyclohexene} \xrightarrow[CF_3CO_2NHEt_3]{(CF_2CO)_2O,\ 90\%} \text{2-hydroxycyclohexyl trifluoroacetate (OCOCF}_3\text{, HO)} \xrightarrow{HCl,\ MeOH} \text{cyclohexane-1,2-diol (OH, HO)} \quad 82\%$$ **[607]**

(364) [608]

$$(CH_3)_2CHCH_2COCH_3 \xrightarrow[Na_2HPO_4,\ CH_2Cl_2,\ reflux]{(CF_3CO)_2O,\ 90\%\ H_2O_2} (CH_3)_2CHCH_2OCOCH_3 \quad 84\%$$

(365) [610]

$$C_6H_5COC(CH_3)_3 \longrightarrow \underset{70\%}{C_6H_5COOC(CH_3)_3} + \underset{2\%}{C_6H_5OCOC(CH_3)_3}$$

$$C_6H_5COCH_2C(CH_3)_3 \longrightarrow \underset{9\%}{C_6H_5COOCH_2C(CH_3)_3} + \underset{82\%}{C_6H_5OCOCH_2C(CH_3)_3}$$

**Table 51. Migration Aptitudes of Alkyls in the Oxidation of Phenyl Alkyl Ketones by Trifluoroperacetic Acid [610]**

| Starting ketone $C_6H_5CO$—R | Yield, % | Composition of the product, % | | Migratory ratio alkyl/phenyl |
|---|---|---|---|---|
| | | $C_6H_5OCOR$ | $C_6H_5COOR$ | |
| $CH_3$ | 90 | 90 | 0 | Very small |
| $C_2H_5$ | 93 | 87 | 6 | 0.07 |
| $C_3H_7$ | 91 | 85 | 6 | 0.07 |
| $(CH_3)_2CH$ | 96 | 33 | 63 | 1.9 |
| $(CH_3)_3C$ | 90 (11% recovered) | 2 | 77 | 39 |
| $(CH_3)_3CCH_2$ | 97 (4% recovered) | 84 | 9 | 0.1 |
| *cyclo*-$C_5H_9$ | 92 | 44 | 48 | 1.1 |
| *cyclo*-$C_6H_{11}$ | 100 | 25 | 75 | 3.0 |
| $C_6H_5CH_2$ | 90 | 39 | 51 | 1.3 |

# References


1. Miller, H. C.: *Chem. & Eng. News* 1949, **27,** 3854; Reinhardt, C. F., Hume, W. G., Linch, A. L., and Wetherhold, J. M.: *J. Chem. Educ.* 1969, **46,** A171.
2. Pattison, F. L. M., and Peters R. A.: "Monofluoro Aliphatic Compounds," in Eichler, O., Farah, A., Herken, H., and Welch, A. D., Handbook of Experimental Pharmacology, Vol. XX, Springer, Berlin, 1966, p. 387.
3. Saunders, B. C.: Phosphorus and Fluorine. Cambridge University Press, Cambridge, 1957.
4. Clayton, J. W., Jr.: *Fluorine Chem. Rev.* 1967, **1,** 197.
5. Perry, J. H.: Chemical Engineers' Handbook. McGraw-Hill, New York, 1963 (4th Ed.).
6. Maxwell, A. F., Detoro, F. E., and Bigelow, L. A.: *J. Am. Chem. Soc.* 1960, **82,** 5827.
7. Cady, G. H., Grosse, A. V., Barber, E. J., Burger, L. L., and Sheldon, Z. D.: *Ind. Eng. Chem.* 1947, **39,** 290.
8. Barbour, A. K., Barlow, G. B., and Tatlow, J. C.: *J. Appl. Chem.* (*London*) 1952, **2,** 127.
9. Kauck, E. A., and Diesslin, A. R.: *Ind. Eng. Chem.* 1951, **43,** 2332.
10. Leech, H. R.: *Chem. & Ind.* (*London*) **1960,** 242.
11. Weinmayr, V.: *J. Am. Chem. Soc.* 1955, **77,** 1762.
12. Munter, P. A., Aepli, O. T., and Kossatz, R. A.: *Ind. Eng. Chem.* 1947, **39,** 427.
13. Rudge, A. J.: The Manufacture and Use of Fluorine and Its Compounds. Oxford University Press, London, 1962.
14. Bergmann, E. D., and Shahak, I.: *J. Chem. Soc.* **1959,** 1418.
15. Buckle, F. J., Pattison, F. L. M., and Saunders, B. C.: *J. Chem. Soc.* **1949,** 1471.
16. Vogel, A. I.: *J. Chem. Soc.* **1948,** 644.
17. Ruggli, P., and Caspar, E.: *Helv. Chim. Acta* 1935, **18,** 1414.
18. Nakanishi, S., Myers, T., and Jensen, E. V.: *J. Am. Chem. Soc.* 1955, **77,** 3099.
19. Hoffman, C. J.: *Inorg. Syn.* 1953, **4,** 150.
20. Pacini, H. A., Teach, E. G., Walker, F. H., and Pavlath, A. E.: *Tetrahedron* 1966, **22,** 1747.
21. Wiechert, K.: *Z. anorg. allgem. Chem.* 1950, **261,** 310.
22. Booth, H. S., and Willson, J. S.: *Inorg. Syn.* 1939, **1,** 21; *J. Am. Chem. Soc.* 1935, **57,** 2273.
23. Horicky, M.: Unpublished results.
24. Smith, W. C., Tullock, C. W., Muetterties, E. L., Hasek, W. R., Fawcett, F. S., Engelhardt, V. A., and Coffman, D. D.: *J. Am. Chem. Soc.* 1959, **81,** 3165.
25. Tullock, C. W., Fawcett, F. S., Smith, W. C., and Coffman, D. D.: *J. Am. Chem. Soc.* 1960, **82,** 539.
26. Seel, F., Jonas, H., Riehl, L., and Langer, J.: *Angew. Chem.* 1955, **67,** 32.
27. La Lande, W. A., Jr.: German Pat. 1,026,285 (1956); *Chem. Abstr.* 1960, **54,** 20117*i*.
28. Sheppard, W. A.: *J. Am. Chem. Soc.* 1962, **84,** 3058.
29. Emeleus, H. J., and Wood, J. F.: *J. Chem. Soc.* **1948,** 2183.

30. Fawcett, F. S., Tullock, C. W., and Coffman, D. D.: *J. Am. Chem. Soc.* 1962, **84,** 4275.
31. Ayer, D. E.: U.S. Pat. 3,105,078 (1963); *Chem. Abstr.* 1964, **60,** 427*b*.
32. Bockemüller W.: *Chem. Ber.* 1931, **64,** 522.
33. Allison, J. A. C., and Cady, G. H.: *J. Am. Chem. Soc.* 1959, **81,** 1089.
34. Smith, W. C.: *J. Am. Chem. Soc.* 1960, **82,** 6176.
35. Anon.: *Ind. Eng. Chem.* 1947, **39,** 241.
36. McBee, E. T., and Hodge, E. B.: *Chem. Eng. News* 1952, **30,** 4513.
37. Grosse, A. V., and Linn, C. B.: *J. Org. Chem.* 1938, **3,** 26.
38. Henne, A. L., and Hinkamp, J. B.: *J. Am. Chem. Soc.* 1945, **67,** 1197.
39. Hopff, H., and Valkanas, G.: *Helv. Chim. Acta* 1963, **46,** 1818.
40. Henne, A. L., and Walkes, T. P.: *J. Am. Chem. Soc.* 1946, **68,** 496.
41. Miller, W. T., Jr., Fried, J. H., and Goldwhite, H.: *J. Am. Chem. Soc.* 1960, **82,** 3091.
42. Miller, W. T., Jr., Freedman, M. B., Fried, J. H., and Koch, H. F.: *J. Am. Chem. Soc.* 1961, **83,** 4105.
43. Miller, W. T., Jr., and Burnard, R. J.: *J. Am. Chem. Soc.* 1968, **90,** 7367.
44. Newkirk, A. E.: *J. Am. Chem. Soc.* 1946, **68,** 2467.
45. Grosse, A. V., and Linn, C. B.: *J. Am. Chem. Soc.* 1942, **64,** 2289.
46. Henne, A. L., and Plueddeman, E. P.: *J. Am. Chem. Soc.* 1943, **65,** 587.
47. Barr, D. A., Haszeldine, R. N.: *J. Chem. Soc.* **1955,** 2532.
48. Buckley, G. D., Piggott, H. A., and Welch, A. J. E.: *J. Chem. Soc.* **1945,** 864.
49. Pittman, A. G., and Sharp, D. L.: *J. Org. Chem.* 1966, **31,** 2316.
50. Evans, F. W., Litt, M. H., Weidler-Kubanek, A. M., and Avonda, F. P.: *J. Org. Chem.* 1968, **33,** 1837.
51. Miller, W. T., Jr., Stoffer, J. O., Fuller, G., and Currie, A. C.: *J. Am. Chem. Soc.* 1964, **86,** 51.
52. Henne, A. L., and Waalkes, T. P.: *J. Am. Chem. Soc.* 1945, **67,** 1639.
53. Bissell, E. R., and Fields, D. B.: *J. Org. Chem.* 1964, **29,** 1591.
54. Bornstein, J., and Skarlos, L.: *J. Am. Chem. Soc.* 1968, **90,** 5044.
55. Bornstein, J., Borden, M. R., Nunes, F., and Tarlin, H. J.: *J. Am. Chem. Soc.* 1963, **85,** 1609.
56. Wood, K. R., and Kent, P. W.: *J. Chem. Soc. C*, **1967,** 2422.
57. Carpenter, W.: *J. Org. Chem.* 1966, **31,** 2688.
58. Robson, P., McLoughlin, V. C. R., Hynes, J. B., and Bigelow, L. A.: *J. Am. Chem. Soc.* 1961, **83,** 5010.
59. Ruff, O., and Willenberg, W.: *Chem. Ber.* 1940, **73,** 724.
60. Chambers, W. J., Tullock, C. W., and Coffman, D. D.: *J. Am. Chem. Soc.* 1962, **84,** 2337.
61. Glemser, O., Schröder, H., and Haeseler, H.: *Z. anorg. allgem. Chem.* 1955, **282,** 80.
62. Emeleus, H. J., and Hurst, G. L.: *J. Chem. Soc.* **1964,** 396.
63. Smith, W. C., Tullock, C. W., Smith, R. D., and Engelhardt, V. A.: *J. Am. Chem. Soc.* 1960, **82,** 551.
64. Cuculo, J. A., and Bigelow, L. A.: *J. Am. Chem. Soc.* 1952, **74,** 710.
65. Hückel, W.: *Chem. Abstr.* 1949, **43,** 6793.
66. Nerdel, F.: Naturwiss. 1952, **39,** 209.
67. Bishop, B. C., Hynes, J. B., and Bigelow, L. A.: *J. Am. Chem. Soc.* 1964, **86,** 1827.
68. Sheppard, W. A.: *J. Am. Chem. Soc.* 1965, **87,** 4338.
69. Emeleus, H. J., and Hurst, G. L.: *J. Chem. Soc.* **1962,** 3276.
70. Cady, G. H., and Kellogg, K. B.: U.S. Pat. 2,689,254 (1954); *Chem. Abstr.* 1955, **49,** 11681*i*.
71. Kellogg, K. B., and Cady, G. H.: *J. Am. Chem. Soc.* 1948, **70,** 3986.
72. Clifford, A. F., El-Shamy, H. K., Emeleus, H. J., and Haszeldine, R. N.: *J. Chem. Soc.* **1953,** 2372.

73. Smith, W. C.: *J. Am. Chem. Soc.* 1960, **82,** 6176.
74. Olah, G. A., and Bollinger, J. M.: *J. Am. Chem. Soc.* 1967, **89,** 4744.
75. Bowers, A., Ibanez, L. C., Denot, E., and Becerra, R.: *J. Am. Chem. Soc.* 1960, **82,** 4001.
76. Dean, F. H., and Pattison, F. L. M.: *Can. J. Chem.* 1965, **43,** 2415.
77. Chambers, R. D., Musgrave, W. K. R., and Savory, J.: *J. Chem. Soc.* **1961,** 3779.
78. Hauptschein, M., and Braid, M.: *J. Am. Chem. Soc.* 1961, **83,** 2383.
79. Krespan, C. G.: *J. Org. Chem.* 1962, **27,** 1813.
80. Gould, D. E., Anderson, L. R., Young, D. E., and Fox, W. B.: *J. Am. Chem. Soc.* 1969, **91,** 1310.
81. Smith, R. D., Fawcett, F. S., and Coffman, D. D.: *J. Am. Chem. Soc.* 1962, **84,** 4285.
82. Chambers, R. D., Jackson, J. A., Musgrave, W. K. R., and Storey, R. A.: *J. Chem. Soc. C,* **1968,** 2221.
83. Cady, G. H., Grosse, A. V., Barber, E. J., Burger, L. L., and Sheldon, Z. D.: *Ind. Eng. Chem.* 1947, **39,** 290.
84. Musgrave, W. K. R., and Smith, F.: *J. Chem. Soc.* **1949,** 3021, 3026.
85. Haszeldine, R. N., and Smith, F.: *J. Chem. Soc.* **1950,** 2689, 2787.
86. Fowler, R. D., Burford, W. B., III, Hamilton, J. M., Jr. Sweet, R. G., Weber, C. E., Kasper, J. S., and Litant, I.: *Ind. Eng. Chem.* 1947, **39,** 292.
87. Barlow, G. B., Stacey, M., and Tatlow, J. C.: *J. Chem. Soc.* **1955,** 1749.
88. Kauck, E. A., and Diesslin, A. R.: *Ind. Eng. Chem.* 1951, **43,** 2332.
89. Scholberg, H. M., and Bryce, H. G.: U.S. Pat. 2,717,871 (1955); *Chem. Abstr.* 1955, **49,** 15572*h*.
90. Freeman, J. P.: *J. Am. Chem. Soc.* 1960, **82,** 3869.
91. Inman, C. E., Oesterling, R. E., and Tyczkowski, E. A.: *J. Am. Chem. Soc.* 1958, **80,** 6533.
92. Gershon, H., Renwick, J. A. A., Wynn, W. K., and D'Ascoli, R.: *J. Org. Chem.* 1966, **31,** 916.
93. Magerlein, B. J., Pike, J. E., Jackson, R. W., Vandenberg, G. E., and Kagan, F.: *J. Org. Chem.* 1964, **29,** 2982.
94. Nakanishi, S.: *Steroids* 1963, **2,** 765; 1964, **3,** 337.
95. Tarrant, P., Attaway, J., and Lovelace, A. M.: *J. Am. Chem. Soc.* 1954, **76,** 2343.
96. Simons, J. H., and Lewis, C. J.: *J. Am. Chem. Soc.* 1938, **60,** 492.
97. Smith, F., Stacey, M., Tatlow, J. C., Dawson, J. K., and Thomas, B. R. J.: *J. Appl. Chem.* 1952, **2,** 97.
98. Brown, J. H., Suckling, C. W., and Whalley, W. B.: *J. Chem. Soc.* **1949,** S95.
99. Daudt, H. W., and Youker, M. A.: U.S. Pat. 2,005,705; 2,005,708 (1935); *Chem. Abstr.* 1935, **29,** 5123.
100. Ruh, R. P., and Davis, R. A.: U.S. Pat. 2,745,867 (1956); *Chem. Abstr.* 1957, **51,** 1245*c*.
101. McBee, E. T., Hass, H. B., Frost, L. W., and Welch, Z. D.: *Ind. Eng. Chem.* 1947, **39,** 404.
102. Vecchio, M.: Ital. Pat. 628,607 (1961); *Chem. Abstr.* 1964, **61,** 4213; unpubl. results.
103. Henne, A. L., and Nager, M.: *J. Am. Chem. Soc.* 1951, **73,** 1042.
104. Haszeldine, R. N.: *J. Chem. Soc.* **1953,** 3371.
105. Gavlin, G., and Maguire, R. G.: *J. Org. Chem.* 1956, **21,** 1342.
106. Nesmeyanov, A. N., Kost, V. N., Zakharin, L. I., and Freidlina, R. K.: *Izv. Akad. Nauk SSSR* **1960,** 447; *Chem. Abstr.* 1960, **54,** 22344*d*.
107. Truce, W. E., Birum, G. H., and McBee, E. T.: *J. Am. Chem. Soc.* 1952, **74,** 3594.
108. Miller, C. B., and Woolf, C.: U.S. Pat. 2,803,665 (1957); *Chem. Abstr.* 1958, **52,** 2047*d*.
109. Henne, A. L., and Trott, P.: *J. Am. Chem. Soc.* 1947, **69,** 1820.
110. Hudlicky, M.: Czech. Pat. 101,089 (1961); *Chem. Abstr.* 1963, **58,** 4424*b*.
111. Latif, A.: *J. Indian Chem. Soc.* 1953, **30,** 525.

112. Yagupolskii, L. M., Troitskaya, V. I., and Malichenko, B. F.: *Zh. Obschei Khim.* 1962, **32,** 1832; *Chem. Abstr.* 1963, **58,** 4446*h*.
113. Tannhauser, P., Pratt, R. J., and Jensen, E. V.: *J. Am. Chem. Soc.* 1956, **78,** 2658.
114. Fort, R. C., Jr., and von Schleyer, P. R.: *J. Org. Chem.* 1965, **30,** 789.
115. Tolman, V., and Veres, K.: *Collection Czech. Chem. Commun.* 1963, **28,** 421.
116. Henne, A. L., and Fox, C. J.: *J. Am. Chem. Soc.* 1954, **76,** 479.
117. Kober, E.: *J. Am. Chem. Soc.* 1959, **81,** 4810.
118. Sheppard, W. A., and Harris, J. F., Jr.: *J. Am. Chem. Soc.* 1960, **82,** 5106.
119. Seel, F., Jonas, H., Riehl, L., and Langer, J.: *Angew. Chem.* 1955, **67,** 32.
120. Nesmeyanov, A. N., and Kahn, E. J.: *Chem. Ber.* 1934, **67,** 370.
121. Olah, G., and Pavlath, A.: *Acta Chim. Acad. Sci. Hung.* 1953, **3,** 191.
122. Bacon, J. C., Bradley, C. W., Hoeberg, E. J., Tarrant, P., and Cassaday, J. T.: *J. Am. Chem. Soc.* 1948, **70,** 2653.
123. Bergman, E. D., and Blank, I.: *J. Chem. Soc.* **1953,** 3786.
124. Fried, J. H., and Miller, W. T., Jr.: *J. Am. Chem. Soc.* 1959, **81,** 2078.
125. Hoffmann, F. W.: *J. Am. Chem. Soc.* 1948, **70,** 2596.
126. Finger, G. C., and Kruse, C. W.: *J. Am. Chem. Soc.* 1956, **78,** 6034.
127. Vorozhtsov, N. N., Jr., Platonov, V. E., and Yakobson, G. G.: *Izv. Akad. Nauk SSSR* **1963,** 1524; *Chem. Abstr.* 1963, **59,** 13846*f*.
128. Chambers, R. D., Hutchinson, J., and Musgrave, W. K. R.: *J. Chem. Soc.* **1964,** 3573; *Proc. Chem. Soc.* **1964,** 83.
129. Saunders, B. C., and Stacey, G. J.: *J. Chem. Soc.* **1948,** 695.
130. Vorozhtsov, N. N., Jr., and Yakobson, G. G.: *Zh. Obshchei Khim.* 1961, **31,** 3705.
131. Nakanishi, S., Myers, T., and Jensen, E. V.: *J. Am. Chem. Soc.* 1955, **77,** 3099.
132. Pacini, H. A., Teach, E. G., Walker, F. H., and Pavlath, A. E.: U.S. Pat. 3,287,424 (1966); *Chem. Abstr.* 1967, **66,** 2678; *Tetrahedron* 1966, **22,** 1747.
133. Dahmlos, J.: *Angew. Chem.* 1959, **71,** 274.
134. Suzuki, Z., and Morita, K.: *Bull. Chem. Soc. Japan* 1968, **41,** 1724.
135. Aranda, G., Jullien, J., and Martin, J. A.: *Bull. Soc. Chim. France* **1965,** 1890.
136. Henbest, H. B., and Wrigley, T. I.: *J. Chem. Soc.* **1957,** 4765.
137. Bergmann, E. D., and Shahak, I.: *Chem. & Ind.* (*London*) **1958,** 157.
138. Edgell, W. F., and Parts, L.: *J. Am. Chem. Soc.* 1955, **77,** 4899.
139. Pattison, F. L. M., and Millington, J. E.: *Can. J. Chem.* 1956, **34,** 757.
140. Wiechert, K., Gruenert, C., and Preibisch, H. J.: *Z. Chem.* 1968, **8,** 64.
141. Cooper, K. A., and Hughes, E. D.: *J. Chem. Soc.* **1937,** 1183.
142. Nakanishi, S., Myers, T., and Jensen, E. V.: *J. Am. Chem. Soc.* 1955, **77,** 3099, 5033.
143. Christe, K. O., and Pavlath, A. E.: *J. Org. Chem.* 1966, **31,** 559.
144. Zappel, S.: *Chem. Ber.* 1961, **94,** 873.
145. Antonucci, J. M., and Wall, L. A.: *Trans* 1963, **3,** 225; *Chem. Abstr.* 1963, **59,** 9844*h*.
146. Kobayashi, Y., and Akashi, C.: Jap. Pat. 68 00,686 (1968); *Chem. Abstr.* 1968, **69,** 35292.
147. Kobayashi, Y., Akashi, C., and Morinaga, K.: *Chem. Pharm. Bull.* (*Tokyo*) 1968, **16,** 1784; *Chem. Abstr.* 1969, **70,** 3168.
148. Yarovenko, N. N., and Raksha, M. A.: *Zh. Obshchei Khim.* 1959, **29,** 2159; *Chem. Abstr.* 1960, **54,** 9724*h*.
149. Hudlicky, M.: *Collection Czech. Chem. Commun.* 1966, **31,** 1416.
150. Kopecky, J., Smejkal, J., and Hudlicky, M.: *Chem. Ind.* (*London*) **1969,** 271.
151. Knox, L. H., Velarde, E., Berger, S., Quadriello, D., and Cross, A. D.: *J. Org. Chem.* 1964, **29,** 2187.
152. Fuqua, S. A., Parkhurst, R. M., and Silverstein, R. M.: *Tetrahedron* 1964, **20,** 1625.
153. Martin, D. G., and Kagan, F.: *J. Org. Chem.* 1962, **27,** 3164.
154. Sheppard, W. A.: *J. Am. Chem. Soc.* 1962, **84,** 3058.
155. Kent, P. W., and Wood, K. R.: Brit. Pat. 1,136,075 (1968); *Chem. Abstr.* 1969, **70,** 88124.

156. Hasek, W. R., Smith, W. C., and Engelhardt, V. A.: *J. Am. Chem. Soc.* 1960, **82,** 543.
157. Bergmann, E. D., and Ikan, R.: *Chem. Ind.* (*London*) **1957,** 394.
158. Machleidt, H., Wessendorf, R., and Klockow, M.: *Ann. Chem.* 1963, **667,** 47.
159. Balz, G., and Schiemann, G.: *Chem. Ber.* 1927, **60,** 1186.
160. Roe, A.: *Org. Reactions* 1949, **5,** 193.
161. Finger, G. C., Starr, L. D., Roe, A., and Link, W. J.: *J. Org. Chem.* 1962, **27,** 3965.
162. Ferm, R. L., and VanderWerf, C. A.: *J. Am. Chem. Soc.* 1950, **72,** 4809.
163. Danek, O.: *Collection Czech. Chem. Commun.* 1964, **29,** 730.
164. Beaty, R. D., and Musgrave, W. K. R.: *J. Chem. Soc.* **1952,** 875.
165. Rutherford, K. G., Redmond, W., and Rigamonti, J.: *J. Org. Chem.* 1961, **26,** 5149.
166. Weiblen, D. G.: *in* Fluorine Chemistry, Volume II (Editor: Simons, J. H.). Academic Press, New York, 1954; p. 449.
167. Brown, J. K., and Morgan, K. J.: *in* Advances in Fluorine Chemistry, Volume 4 (Editors: Stacey, M., Tatlow, J. C., and Sharpe, A. G.). Butterworths, London, 1965; p. 253.
168. The Sadtler Standard Spectra. Sadtler Research Laboratories, Philadelphia, Pa.
169. Kakac, B., and Hudlicky, M.: *Talanta* 1962, **9,** 530; *Collection Czech. Chem. Commun.* 1962, **27,** 2616.
170. Catalogue of Mass Spectral Data. A.P.I. Project 44, Carnegie Institute of Technology, Pittsburgh, Pa.
171. Stenhagen, E., Abrahamsson, S., and McLafferty, F. W.: Atlas of Mass Spectral Data. John Wiley and Sons, New York, 1969.
172. Majer, J. R.: *in* Advances of Fluorine Chemistry, Volume 2 (Editors: Stacey, M., Tatlow, J. C., and Sharpe, A. G.). Butterworths, London, 1961; p. 55.
173. Bovey, F. A.: NMR Data Tables for Organic Compounds. Interscience Publishers, New York, 1967.
174. Emsley, J. W., Feeney, J., and Sutcliffe, L. H.: High Resolution Nuclear Magnetic Resonance Spectroscopy, Volume 2. Pergamon Press, Oxford, 1966.
175. Brey, W. L., and Hynes, J. L.: *in* Fluorine Chemistry Reviews, Volume 2 (Editor: Tarrant, P.). Marcel Dekker, New York, 1968, p. 111.
176. Hoffman, C. J.: *in* Fluorine Chemistry Reviews, Volume 2 (Editor: Tarrant, P.). Marcel Dekker, New York, 1968, p. 161.
177. Evans, D. E. M., and Tatlow, J. C.: *in* Vapor Phase Chromatography (Editors: Desty, D. H., and Harbourn, C. L. A.). Butterworths, London, 1956.
178. Percival, W. C.: *Anal. Chem.* 1957, **29,** 20.
179. Lysyj, I., and Newton, P. R.: *Anal. Chem.* 1963, **35,** 90.
180. Pollard, F. H., and Hardy, C. J.: *Anal. Chim. Acta* 1957, **16,** 135.
181. Campbell, R. H., and Gudzinovicz, B. J.: *Anal. Chem.* 1961, **33,** 842.
182. Pappas, W. S., and Million, J. G.: *Anal. Chem.* 1968, **40,** 2176.
183. Weygand, F., Kolb, B., and Kirchner, P.: *Z. Anal. Chem.* 1961, **181,** 396.
184. Rotzsche, H.: *Z. Anal. Chem.* 1960, **175,** 338.
185. Cruickshank, P. A., and Sheehan, J. C.: *Anal. Chem.* 1964, **36,** 1191.
185a. Landault, C., Guichon, G., and Ganausia, J.: *Bull. Soc. Chim. France* **1967,** 3985.
186. Belcher, R., Leonard, M. A., and West, T. S.: *Talanta* 1959, **2,** 93.
187. Stone, I.: *J. Chem. Educ.* 1931, **8,** 347.
188. Körbl, J.: Personal communication.
189. Elving, P. J., Horton, C. A., and Willard, H. H.: *in* Fluorine Chemistry, Volume II (Editor: Simons, J. H.). Academic Press, New York, 1954; p. 51.
190. Taylor, N. F., and Kent, P. W.: *J. Chem. Soc.* **1958,** 872.
191. Willard, H. H., and Winter, O. B.: *Ind. Eng. Chem. Anal. Ed.* 1933, **5,** 7.
192. Lingane, J. J.: *Anal. Chem.* 1967, **39,** 881; 1968, **40,** 935.
193. Covington, A. K.: *Chem. in Britain* 1969, **5,** 388.
194. Durst, R. A.: *Anal. Chem.* 1968, **40,** 931.

195. Ma, T. S.: *Anal. Chem.* 1958, **30,** 1557; *Microchem. J.* 1958, **2,** 91.
196. Ma, T. S., and Qwirtsman, J.: *Anal. Chem.* 1957, **29,** 140.
197. Johncock, P., Musgrave, W. K. R., and Wiper, A.: *Analyst* 1959, **84,** 245.
198. Wickbold, R.: *Angew. Chem.* 1952, **64,** 133; 1954, **66,** 173.
199. Ehrenberger, F.: *Mikrochim. Acta.* 1959, **2,** 192.
200. Horacek, J., and Pechanec, V.: *Mikrochim. Acta.* **1966,** 17.
201. Steyermark, A., Kaup, R. R., Petras, D. A., and Bass, E. A.: *Microchem. J.* 1959, **3,** 523.
202. Gelman, N. E., Korshun, M. O., and Sheveleva, N. S.: *Zh. Anal. Khim.* 1957, **18,** 526; *Chem. Abstr.* 1958, **52,** 1853.
203. Nikolaev, N. S.: *Chem. Age* (*N. Y.*) 1946, **54,** 309; *Chem. Abstr.* 1946, **40,** 4316.
204. Gagnon, J. G., and Olson, P. B.: *Anal. Chem.* 1968, **40,** 1856.
205. Horacek, J., and Körbl, and J.: *Collect. Czech. Chem. Commun.* 1959, **24,** 286.
206. Otto, K., Uhlir, M., and Pesa, J.: *Chem. Prum.* 1959, **9,** 587; *Chem. Abstr.* 1960, **54,** 3056.
207. Pennington, W. A.: *Anal. Chem.* 1949, **21,** 766.
208. Isceon. Imperial Smelting Corporation, 1957.
209. Henne, A. L., and Stewart, J. J.: *J. Am. Chem. Soc.* 1955, **77,** 1901.
210. Henne, A. L., and Francis, W. C.: *J. Am. Chem. Soc.* 1953, **75,** 991.
211. Filler, R., and Schure, R. M.: *J. Org. Chem.* 1967, **32,** 1217.
212. Haszeldine, R. N.: *J. Am. Chem. Soc.* 1953, **75,** 991; (footnote 4).
213. Forbes, E. J., Richardson, R. D., Stacey, M., and Tatlow, J. C.: *J. Chem. Soc.* **1959,** 2019.
214. Henne, A. L., and Fox, C. J.: *J. Am. Chem. Soc.* 1951, **73,** 2323; 1953, **75,** 5750.
215. Emeleus, H. J., Haszeldine, R. N., and Paul, R. C.: *J. Chem. Soc.* **1955,** 563.
216. Streitwieser, A., Jr., and Holtz, D.: *J. Am. Chem. Soc.* 1967, **89,** 692.
217. Saunders, B. C.: Phosphorus and Fluorine. Some Aspects of the Chemistry and Toxic Action of Their Organic Compounds. Cambridge Uniersity Press, Cambridge, 1957.
218. Pattison, F. L. M.: Toxic Aliphatic Fluorine Compounds. Elsevier, Amsterdam, 1959.
219. Peters, R.: *Endeavour* 1954, **13,** 147.
220. Oelrichs, P. B., and McEwan, T.: *Nature* 1961, **190,** 808.
221. Peters, R. A., Hall, R. J., Ward, P. F. V., and Sheppard, N.: *Biochem. J.* 1960, **77,** 17, 22.
222. Carpenter, C. P., Smyth, H. F., Jr., and Pozzani, U. C.: *J. Ind. Hyg. Toxicol.* 1949, **31,** 343.
223. Clayton, J. W., Jr.: *Fluorine Chem. Rev.* 1967, **1,** 197.
224. Hamilton, J. M., Jr.: *in* Advances in Fluorine Chemistry, Volume 3 (Editors: Stacey, M., Tatlow, J. C., and Sharpe, A. G.). Butterworths, London, 1963; p. 117.
225. Bryce, H. G.: *in* Fluorine Chemistry, Volume 5 (Editor: Simons, J. H.). Academic Press, New York, 1964; p. 295.
226. Midgley, T., Jr., and Henne, A. L.: *Ind. Eng. Chem.* 1930, **22,** 542.
227. Daudt, H. W., and Youker, M. A.: U.S. Pat. 2,005,705; 2,005,708 (1935); *Chem. Abstr.* 1935, **29,** 5123.
228. McBee, E. T., Hass, H.B., Frost, L. W., and Welch, Z.D.: *Ind. Eng. Chem.* 1947, **39,** 404.
229. Farbwerke Hoechst: Fr. Pat. 1,343,392 (1963); *Chem. Abstr.* 1964, **60,** 9147b.
230. Vecchio, M.: Ital. Pat. 628,607 (1961); *Chem. Abstr.* 1964, **61,** 4213*f.*
231. Barbour, A. K.: *Chem. & Ind.* (*London*) **1961,** 958.
232. Rosenberg, H., and Mosteller, J. C.: *Ind. Eng. Chem.* 1953, **45,** 2283.
233. Plunkett, R. J.: U.S. Pat. 2,230,654 (1941); *Chem. Abstr.* 1941, **34,** 3365.
234. Park, J. D., Benning, A. F., Downing, F. B., Laucius, J. F., and McHarness, R. C.: *Ind. Eng. Chem.* 1947, **39,** 354.

235. Belmore, E. A., Ewalt, W. M., and Wojcik, B. H.: *Ind. Eng. Chem.* 1947, **39,** 338.
236. Slesser, O., and Schram, S. R.: Preparation, Properties, and Technology of Fluorine and Organic Fluoro Compounds. McGraw-Hill Book Co., New York, 1959; p. 593.
237. Sianesi, D., and Fontanelli, R.: *Ann. Chim.* (*Rome*) 1965, **55,** 850.
238. Newkirk, A. E.: *J. Am. Chem. Soc.* 1946, **68,** 2467.
239. McBee, E. T., Hill, H. M., and Bachman, G. B.: *Ind. Eng. Chem.* 1949, **41,** 70.
240. Codding, D. W., Reid, T. S., Ahlbrecht, A. H., Smith, G. H., Jr., and Husted, D. R.: *J. Polymer Sci.* 1955, **15,** 515.
241. Pierce, O. R., Holbrook, G. W., Johannson, O. K., Saylor, J. C., and Brown, E. D.: *Ind. Eng. Chem.* 1960, **52,** 783.
242. Henry, M. C., Griffis, C. B., and Stump, E. C.: *Fluorine Chem. Rev.* 1967, **1,** 1.
243. Miller, W. T., Jr., Dittman, A. L., Ehrenfeld, R. L., and Prober, M.: *Ind. Eng. Chem.* 1947, **39,** 333.
244. Hamilton, J. M., Jr.: *Ind. Eng. Chem.* 1953, **45,** 1347.
245. Elliott, J. R., Myers, R. L., and Roedel, G. F.: *Ind. Eng. Chem.* 1953, **45,** 1786.
246. Hodge, H. C., Smith, F. A., and Chen, P. S.: *in* Fluorine Chemistry, Volume III (Editor: Simons, J. H.). Academic Press, New York, 1963.
247. Hine, J., and Ghirardelli, R. G.: *J. Org. Chem.* 1958, **23,** 1550.
248. Lindsey, R. V., Jr., England, D. C., and Andreades, S.: *Chem. Eng. News* 1961, **39** (37), 56.; *J. Am. Chem. Soc.* 1961, **83,** 4670.
249. Hauptschein, M., and Brown, R. A.: *J. Am. Chem. Soc.* 1955, **77,** 4930.
250. Hudlicky, M.: *Collection Czech. Chem. Commun.* 1961, **26,** 3140; unpublished results.
251. Simmons, H. E., and Wiley, D. W.: *J. Am. Chem. Soc.* 1960, **82,** 2288.
252. Braendlin, H. P., and McBee, E. T.: *in* Advances in Fluorine Chemistry, Volume 3 (Editors: Stacey, M., Tatlow, J. C., and Sharpe, A. G.). Butterworths, London, 1963; p. 1.
253. Walborsky, H. M., Baum, M., and Loncrini, D. F.: *J. Am. Chem. Soc.* 1955, **77,** 3637.
254. McBee, E. T.: Intern. Symposium on Fluorine Chem., Birmingham, 1959.
255. Henne, A. L., and Kraus, D.: Thesis, Ohio State University, 1953.
256. Miller, W. T., Jr., Fried, J. H., and Goldwhite, H.: *J. Am. Chem. Soc.* 1960, **82,** 3091.
257. Forbes, E. J., Richardson, R. D., Stacey, M., and Tatlow, J. C.: *J. Chem. Soc.* **1959,** 2019.
258. Miller, J., and Yeung, H. W.: *Aust. J. Chem.* 1967, **20,** 379.
259. Chambers, R. D., Jackson, J. A., Musgrave, W. K. R., and Storey, R. A.: *J. Chem. Soc. C* **1968,** 2221.
260. Smith, R. D., Fawcett, F. S., and Coffman, D. D.: *J. Am. Chem. Soc.* 1962, **84,** 4285.
261. Haszeldine, R. N.: *J. Chem. Soc.* **1953,** 1748.
262. Henne, A. L., and Kaye, S.: *J. Am. Chem. Soc.* 1950, **72,** 3369.
263. Knunyants, I. L., Pervova, E. Y., and Tyuleneva, V. V.: *Izv. Akad. Nauk SSSR* **1956,** 843; *Chem. Abstr.* 1957, **51,** 1814*g*.
264. Chen Tsin-Yun, Gambaryan, N. P., and Knunyants, I. L.: *Doklady Adad. Nauk SSSR* 1960, **133,** 1113; *Chem. Abstr.* 1960, **54,** 24385*g*.
265. Simons, J. H., and Ramler, E. O.: *J. Am. Chem. Soc.* 1943, **65,** 389.
266. Park, J. D., Sharah, M. L., and Lacher, J. R.: *J. Am. Chem. Soc.* 1949, **71,** 2339.
267. Englund, B.: *Org. Syn.* 1954, **34,** 16; *Coll. Vol.* 1963, **4,** 184.
268. Olah, G., Pavlath, A., Kuhn, I., and Varsanyi, G.: *J. Chem. Soc.* **1957,** 1823.
269. Suschitzky, H.: *J. Chem. Soc.* **1955,** 4026.
270. Illuminati, G., and Marino, G.: *J. Am. Chem. Soc.* 1956, **78,** 4975.
271. Haszeldine, R. N., and Steele, B. R.: *J. Chem. Soc.* **1957,** 2193, 2800.
272. Park, J. D., Seffl, R. J., and Lacher, J. R.: *J. Am. Chem. Soc.* 1956, **78,** 59.
273. Henne, A. L., and Ruh, R. P.: *J. Am. Chem. Soc.* 1949, **69,** 279.

274. Parker, R. E.: *in* Advances in Fluorine Chemistry, Volume 3 (Editors: Stacey, M., Tatlow, J. C., and Sharpe, A. G.). Butterworths, London, 1963; p. 65.
275. Hudlicky, M., Kraus, E., Körbl, J., and Cech, M.: *Collection Czech. Chem. Comm.* 1969, **34,** 833.
276. Tarrant, P., and Brown, H. C.: *J. Am. Chem. Soc.* 1951, **73,** 1781.
277. Pruett, R. L., Barr, J. T., Rapp, K. E., Bahner, C. T., Gibson, J. D., and Lafferty, R. H., Jr.: *J. Am. Chem. Soc.* 1950, **72,** 3646.
278. Tiers, G. V. D.: *J. Am. Chem. Soc.* 1955, **77,** 4837, 6703, 6704.
279. LaZerte, J. D., Hals, L. J., Reid, R. S., and Smith, G. H.: *J. Am. Chem. Soc.* 1953, **75,** 4525.
280. Bornstein, J., and Borden, M. R.: *Chem. & Ind. (London)* **1958,** 441.
281. Carpenter, W.: *J. Org. Chem.* 1966, **31,** 2688.
282. Miller, W. T., Jr., Fager, E. W., and Griswald, P. H.: *J. Am. Chem. Soc.* 1950, **72,** 705.
283. Oelschläger, H., and Schmersahl, P.: *Arch. Pharm.* 1963, **296,** 324.
284. Knunyants, I. L., Mysov, E. I., and Krasuskaya, M. P.: *Izv. Akad. Nauk SSSR* **1958,** 906; *Chem. Abstr.* 1959, **53,** 1102*b*.
285. Swarts, F.: *Bull. acad. roy. Belgique* **1920,** 399; *Chem. Zentr.* 1921, **III,** 32.
286. Pattison, F. L. M., and Saunders, B. C.: *J. Chem. Soc.* **1949,** 2745.
287. Pummer, W. J., Wall, L. A., and Florin, R. E.: *Chem. Eng. News* 1958, **36**(48), 42, 44.
288. Robson, P., Stacey, M., Stephens, R., and Tatlow, J. C.: *J. Chem. Soc.* **1960,** 4754.
289. Chambers, R. D., Drakesmith, F. G., and Musgrave, W. K. R.: *J. Chem. Soc.* **1965,** 5045.
290. Haszeldine, R. N., and Osborne, J. E.: *J. Chem. Soc.* **1956,** 61.
291. Feast, W. J., Perry, D. R. A., and Stephens, R.: *Tetrahedron* 1966, **22,** 433.
292. Banks, R. E., Haszeldine, R. N., Latham, J. V., and Young, I. M.: *J. Chem. Soc.* **1965,** 594.
293. Brooke, G. M., Burdon, J., and Tatlow, J. C.: *J. Chem. Soc.* **1962,** 3253.
294. Brooke, G. M., Chambers, R. D., Heyes, J., and Musgrave, W. K. R.: *Proc. Chem. Soc.* **1963,** 213.
295. Dannley, R. L., Taborsky, R. G., and Lukin, M.: *J. Org. Chem.* 1956, **21,** 1318.
296. Dannley, R. L., and Taborsky, R. G.: *J. Org. Chem.* 1957, **22,** 77.
297. Meier, R., and Böhler, F.: *Chem. Ber.* 1957, **90,** 2344.
298. Fear, E. J. P., Thrower, J., and Veitch, J.: *J. Appl. Chem.* 1955, **5,** 589.
299. Hudlicky, M., Lejhancova, I.: *Collection Czech. Chem. Comm.* 1965, **30,** 2491.
300. McGinty, R. L.: U.S. Pat. 3,082,263 (1963); *Chem. Abstr.* 1963, **59,** 8592*c*.
301. Madai, H., and Müller, R.: *J. prakt. Chem.* 1963, **19,** 83.
302. Allen, D. R.: *J. Org. Chem.* 1961, **26,** 923.
303. Haszeldine, R. N., and Nyman, F.: *J. Chem. Soc.* **1959,** 420.
304. Farbwerke Hoechst, A. G.: Neth. Appl. 6,611,128 (1967); *Chem. Abstr.* 1967, **67,** 5032.
305. Müller, W., and Walaschewsky, E.: Ger. Pat. 947,364 (1956); *Chem. Abstr.* 1959, **53,** 4299*b*.
306. Hurka, V. R.: U. S. Pat. 2,676,983 (1954); *Chem. Abstr.* 1955, **49,** 5510*g*.
307. Moore, L. O., and Clark, J. W.: *J. Org. Chem.* 1965, **30,** 2472.
308. Haszeldine, R. N.: *J. Chem. Soc.* **1953,** 3565.
309. Pearlson, W. H., and Hals, L. J.: U.S. Pat. 2,617,836 (1952); *Chem. Abstr.* 1953, **47,** 8770*e*.
310. Yakubovich, A. J., Gogol, V., Borzova, I.: *Zh. Prikl. Khim.* 1959, **32,** 451; *Chem. Abstr.* 1959, **53,** 13045*i*.
311. Henne, A. L.: U.S. Pat. 2,371,757 (1945); *Chem. Abstr.* 1945, **39,** 3307; Henne, A. L., and Trott, P.: *J. Am. Chem. Soc.* 1947, **69,** 1820.
312. Henne, A. L., Alderson, T., and Newman, M. S.: *J. Am. Chem. Soc.* 1945, **67,** 918.
313. Burdon, J., and Tatlow, J. C.: *J. Appl. Chem.* 1958, **8,** 293.
314. Hudlicky, M.: *Collection Czech. Chem. Comm.* 1960, **25,** 1199.

315. Korshak, V. V., and Kolesnikov, G. S.: *Syntesi org. soedinenii* 1952, **2,** 140; *Chem. Abstr.* 1954, **48,** 635*f.*
316. Hopff, H., and Valkanas, G.: *J. Org. Chem.* 1962, **27,** 2923; *J. Chem. Soc.* **1963,** 3475.
317. Wächter, R.: *Angew. Chem.* 1955, **67,** 305.
318. Haworth, W. N., and Stacey, M.: U.S. Pat. 2,476,490 (1949); *Chem. Abstr.* 1950, **44,** 654*i.*
319. Gale, D. M., Middleton, W. J., and Krespan, C. G.: *J. Am. Chem. Soc.* 1965, **87,** 657.
320. Malatesta, P., and D'Atri, B.: *Ricerca sci.* 1952, **22,** 1589.
321. Millington, J. E., Brown, G. M., and Pattison, F. L. M.: *J. Am. Chem. Soc.* 1956, **78,** 3846.
322. Saunders, B. C., Stacey, G. J., and Wilding, I. G. E.: *J. Chem. Soc.* **1949,** 773.
323. Pattison, F. L. M., Stothers, J. B., and Woolford, R. G.: *J. Am. Chem. Soc.* 1956, **78,** 2255.
324. Evans, D. E. M., Feast, W. J., Stephens, R., and Tatlow, J. C.: *J. Chem. Soc.* **1963,** 4828.
325. Coffman, D. D., Raasch, M. S., Rigby, G. W., Barrick, P. L., and Hanford, W. E.: *J. Org. Chem.* 1949, **14,** 747.
326. Haszeldine, R. N., and Leedham, K.: *J. Chem. Soc.* **1953,** 1548.
327. Twelves, R. R., and Weinmayr, V.: U.S. Pat. 3,076,041 (1963); *Chem. Abstr.* 1963, **58,** 13792*d.*
328. Chambers, R. D., Musgrave, W. K. R., and Savory, J.: *J. Chem. Soc.* **1961,** 3779.
329. Hauptschein, M., and Braid, M.: *J. Am. Chem. Soc.* 1961, **83,** 2383.
330. Henne, A. L., and Hinkamp, J. B.: *J. Am. Chem. Soc.* 1945, **67,** 1197.
331. Olah, G., Pavlath, A., and Varsanyi, G.: *J. Chem. Soc.* **1957,** 1823.
332. Varma, P. S., Venkataraman, K. S., and Nilkantiah, P. M.: *J. Ind. Chem. Soc.* 1944, **21,** 112.
333. Simons, J. H., and Ramler, E. O.: *J. Am. Chem. Soc.* 1943, **65,** 389.
334. Benkeser, R. A., and Severson, R. G.: *J. Am. Chem. Soc.* 1951, **73,** 1353.
335. Nield, E., Stephens, R., and Tatlow, J. C.: *J. Chem. Soc.* **1959,** 166.
336. Hauptschein, M., Stokes, C. S., and Grosse, A. V.: *J. Am. Chem. Soc.* 1952, **74,** 848.
337. Barlow, G. B., Stacey, M., and Tatlow, J. C.: *J. Chem. Soc.* **1955,** 1749.
338. Knunyants, I. L., Shokina, V. V., and Kuleshova, N. D.: *Izv. Akad. Nauk SSSR* **1960,** 1693; *Chem. Abstr.* 1961, **55,** 9254*g.*
339. Park, J. D., Sharrah, M. L., and Lacher, J. R.: *J. Am. Chem. Soc.* 1949, **71,** 2339.
340. Haszeldine, R. N., and Steele, B. R.: *J. Chem. Soc.* **1954,** 3747.
341. Zahalka, J., and Hudlicky, M.: Czech. Pat. 116,341 (1965); *Chem. Abstr.* 1966, **65,** 2125*a.*
342. Hudlicky, M.: Czech. Pat. 116,813 (1965); *Chem. Abstr.* 1966, **65,** 13540*h.*
343. Haszeldine, R. N.: *J. Chem. Soc.* **1952,** 3490.
344. Gilman, H., and Jones, R. G.: *J. Am. Chem. Soc.* 1943, **65,** 2037.
345. Solomon, W. C., Dee, L. A., and Schults, D. W.: *J. Org. Chem.* 1966, **31,** 1551.
346. Tiers, G. V. D.: *J. Am. Chem. Soc.* 1955, **77,** 4837, 6703, 6704.
347. Knunyants, I. L., Dyatkin, B. L., Fokin, A. V., and Komarov, V. A.: *Izv. Akad. Nauk SSSR* **1964,** 1425; *Chem. Abstr.* 1966, **64,** 19394*b.*
348. Scribner, R. M.: *J. Org. Chem.* 1964, **29,** 284.
349. Boschan, R.: *J. Org. Chem.* 1960, **25,** 1450.
350. Zahn, H., and Würz, A.: *Angew. Chem.* 1951, **63,** 147.
351. Coe, P. L., Jukes, A. E., and Tatlow, J. C.: *J. Chem. Soc. C,* **1966,** 2323.
352. Shteingarts, V. D., Osina, O. I., Yakobson, G. G., and Vorozhtsov, N. N., Jr.: *Zh. Vses. Khim. Obshchestva im. D.I. Mendeleeva* 1966, **11,** 115; *Chem. Abstr.* 1966, **64,** 17506*a.*
353. Jander, J., and Haszeldine, R. N.: *J. Chem. Soc.* **1954,** 912.

354. Taylor, C. W., Brice, T. J., and Wear, R. L.: *J. Org. Chem.* 1962, **27,** 1064.
355. Gilman, H., and Jones, R. G.: *J. Am. Chem. Soc.* 1943, **65,** 1458.
356. Forbes, E. J., Richardson, R. D., and Tatlow, J. C.: *Chem. Ind. (London)* **1958,** 630.
357. Nield, E., Stephens, R., and Tatlow, J. C.: *J. Chem. Soc.* **1959,** 166.
358. Paleta, O., and Posta, A.: *Collection Czech. Chem. Commun.* 1966, **31,** 3584; 1967, **32,** 1427.
359. Knunyants, I. L., Sterlin, R. N., Pinkina, L. M., and Dyatkin, B. L.: *Izv. Acad. Nauk SSSR* **1958,** 296; *Chem. Abstr.* 1958, **52,** 12754*e*.
360. Johnson, L. V., Smith, F., Stacey, M., and Tatlow, J. C.: *J. Chem. Soc.* **1952,** 4710.
361. Olah, G., Pavlath, A., and Kuhn, I.: *Acta Chim. Acad. Sci. Hung.* 1955, **7,** 85.
262. Hamada, M., and Nagasawa, S.: *Botyu-Kyaku* 1956, **21,** 4; *Chem. Abstr.* 1957, **51,** 3518*g*.
363. Buu-Hoi, N. P., Hoan, N., and Jacquignon, P.: *Rec. Trav. Chim.* 1949, **68,** 781.
364. Beckert, W. F., and Lowe, J. U., Jr.: *J. Org. Chem.* 1967, **32,** 582.
365. Olah, G. A., and Kuhn, S. J.: *J. Org. Chem.* 1964, **29,** 2317.
366. Bergmann, F., and Kalmus, A.: *J. Am. Chem. Soc.* 1954, **76,** 4137.
367. Prober, M.: *J. Am. Chem. Soc.* 1953, **75,** 968.
368. Cohen, S. G., Wolosinski, H. T., and Scheuer, P. J.: *J. Am. Chem. Soc.* 1949. **71,** 3439.
369. Vorozhtsov, N. N., Jr., Barkhash, V. A., and Anichkina, S. A.: *Zh. Organ. Khim.* 1966, **2,** 1903; *Chem. Abstr.* 1967, **66,** 6122.
370. Inman, C. E., Oesterling, R. E., and Tyczkowski, E. A.: *J. Am. Chem. Soc.* 1958, **80,** 5286.
371. Codding, D. W., Reid, T. S., Ahlbrecht, A. H., Smith, G. H., Jr., and Husted, D. R.: *J. Polymer Sci.* 1955, **15,** 515.
372. Henne, A. L., Alderson, T., and Newman, M. S.: *J. Am. Chem. Soc.* 1945, **67,** 918.
373. Reid, J. C.: *J. Am. Chem. Soc.* 1947, **69,** 2069.
374. Benoiton, L., Rydon, H. N., and Willett, J. E.: *Chem. Ind. (London)* **1960,** 1060.
375. Bacon, J. C., Bradley, C. W., Hoeberg, E. I., Tarrant, P., and Cassaday, J. T.: *J. Am. Chem. Soc.* 1948, **70,** 2653.
376. Prikryl, J.: Personal communication.
377. Pattison, F. L. M., Stothers, J. B., and Woolford, R. G.: *J. Am. Chem. Soc.* 1956, **78,** 2255.
378. Zarubinskii, G. M., Koltsov, A. I., Orestova, V. A., and Danilov, S. N.: *Zh. Obshchei Khim.* 1965, **35,** 1620; *Chem. Abstr.* 1965, **63,** 17883*b*.
379. Grindahl, G. A., Bajzer, W. X., and Pierce, O. R.: *J. Org. Chem.* 1967, **32,** 603.
380. Stockel, R. F., Beachem, M. T., and Megson, F. H.: *J. Org. Chem.* 1965, **30,** 1629.
381. Goldstein, H., and Giddey, A.: *Helv. Chim. Acta* 1954, **37,** 1121.
382. Birchall, J. M., and Haszeldine, R. N.: *J. Chem. Soc.* **1959,** 13.
383. Chambers, R. D., Hutchinson, J., and Musgrave, W. K. R.: *J. Chem. Soc.* **1964,** 5634.
384. Saunders, B. C., and Stacey, G. J.: *J. Chem. Soc.* **1948,** 695.
385. Silversmith, E. F., and Roberts, J. D.: *J. Am. Chem. Soc.* 1958, **80,** 4083.
386. Roberts, J. D., Kline, G. B., and Simmons, H. E., Jr.: *J. Am. Chem. Soc.* 1953, **75,** 4765.
387. Sweeney, R. F., Veldhuis, B., Gilbert, E. E., Anello, L. G., Dubois, R. J., and Cunningham, W. J.: *J. Org. Chem.* 1966, **31,** 3174.
388. Yarovenko, N. N., and Raksha, M. A.: *Zh. Obschei Khim.* 1959, **29,** 2159; *Chem. Abstr.* 1960, **54,** 9724*h*.
389. Young, J. A., and Tarrant, P.: *J. Am. Chem. Soc.* 1949, **71,** 2432.
390. Tarrant, P., and Brown, H. C.: *J. Am. Chem. Soc.* 1951, **73,** 5831.
391. Buxton, M. W., Stacey, M., and Tatlow, J. C.: *J. Chem. Soc.* **1954,** 366.
392. Swarts, F.: *Bull. acad. roy. Belg.* **1920,** 389; *Chem. Zentr.* 1921, **III,** 32.
393. LeFave, G. M.: *J. Am. Chem. Soc.* 1949, **71,** 4148.

394. Simons, J. H., and McArthur, R. E.: *Ind. Eng. Chem.* 1947, **39,** 366.
395. Minnesota Mining & Manufacturing Co.: Br. Pat. 737,164 (1955); *Chem. Abstr.* 1956, **50,** 13987*d*.
396. McBee, E. T., and Rapkin, E.: *J. Am. Chem. Soc.* 1951, **73,** 1366.
397. Simons, J. H., and Ramler, E. O.: *J. Am. Chem. Soc.* 1943, **65,** 389.
398. Bergman, E.: *J. Org. Chem.* 1958, **23,** 476.
399. Tronov, B., and Krüger, E.: *Zh. Russ. Fyz. Khim. Obshchestva* 1926, **58,** 1270; *Chem. Abstr.* 1927, **12,** 3887.
400. Maguire, M. H., and Shaw, G.: *J. Chem. Soc.* **1957,** 2713.
401. Bronnert, D. L. E., and Saunders, B. C.: *Tetrahedron* 1960, **10,** 160.
402. Tarrant, P., and Brown, H. C.: *J. Am. Chem. Soc.* 1951, **73,** 5831.
403. Park, J. D., Wilson, L. H., and Lacher, J. R.: *J. Org. Chem.* 1963, **28,** 1008.
404. Shepard, R. A., Lessoff, H., Domijan, J. D., Hilton, D. B., and Finnegan, T. F.: *J. Org. Chem.* 1958, **23,** 2011.
405. Park, J. D., Dick, J. R., and Adams, J. H.: *J. Org. Chem.* 1965, **30,** 400.
406. Howell, W. C., Millington, J. E., and Pattison, F. L. M.: *J. Am. Chem. Soc.* 1956, **78,** 3843.
407. Hine, J., and Porter, J. J.: *J. Am. Chem. Soc.* 1957, **79,** 5493.
408. Rapp, K. E., Pruett, R. L., Barr, J. T., Bahner, C. T., Gibson, J. D., and Lafferty, R. H., Jr.: *J. Am. Chem. Soc.* 1950, **72,** 3642.
409. Walborsky, H. M., and Baum, M. E.: *J. Org. Chem.* 1956, **21,** 538; Walborsky, H. M., Baum, M. E., and Loncrini, D. F.: *J. Am. Chem. Soc.* 1955, **77,** 3637.
410. Knunyants, I. L., German, L. S., and Dyatkin, B. L.: *Izv. Akad. Nauk. SSSR* **1956,** 1353; *Chem. Abstr.* 1957, **51,** 8037*f*.
411. McBee, E. T., Turner, J. J., Morton, C. J., and Stefani, A. P.: *J. Org. Chem.* 1965, **30,** 3698.
412. Frank, A. W.: *J. Org. Chem.* 1966, **31,** 1917.
413. Pattison, F. L. M., Cott, W. J., Howell, W. C., and White, R. W.: *J. Am. Chem. Soc.* 1956, **78,** 3484.
414. Pattison, F. L. M., and Norman, J. J.: *J. Am. Chem. Soc.* 1957, **79,** 2311.
415. Bergmann, E. D., and Szinai, S.: *J. Chem. Soc.* **1956,** 1521.
416. Knunyants, I. L., Shokina, V. V., and Khrlakyan, S. P.: USSR Pat. 170,932 (1965); *Chem. Abstr.* 1965, **63,** 9893*a*.
417. Bevan, C. W. L.: *J. Chem. Soc.* **1951,** 2340.
418. Wall, L. A., Pummer, W. J., Fearn, J. F., and Antonucci, J. M.: *J. Res. Natl. Bur. Std.* 1963, **67A,** 481; *Chem. Abstr.* 1964, **60,** 9170*b*.
419. Forbes, E. J., Richardson, R. D., Stacey, M., and Tatlow, J. C.: *J. Chem. Soc.* **1959,** 2019.
420. Pummer, W. J. and Wall, L. A.: *Science* 1958, **127,** 643.
421. Birchall, J. M., and Haszeldine, R. N.: *J. Chem. Soc.* **1959,** 13.
422. Banks, R. E., Field, D. S., and Haszeldine, R. N.: *J. Chem. Soc. C,* **1967,** 1822.
423. Robson, P., Stacey, M., Stephens, R., and Tatlow, J. C.: *J. Chem. Soc.* **1960,** 4754.
424. Robson, P., Smith, T. A., Stephens, R., and Tatlow, J. C.: *J. Chem. Soc.* **1963,** 3692.
425. Burdon, J., Damodaran, V. A., and Tatlow, J. C.: *J. Chem. Soc.* **1964,** 763.
426. Yakobson, G. G., Shteingarts, V. D., Furin, G. G., and Vorozhtsov, N. N., Jr.: *Zh. Obshchei Khim.* 1964, **34,** 3514; *Chem. Abstr.* 1965, **62,** 2724*e*.
427. Holland, D. G., Moore, G. J., and Tamborski, C.: *J. Org. Chem.* 1964, **29,** 1562.
428. Chambers, R. D., Hutchinson, J., and Musgrave, W. K. R.: *J. Chem. Soc.* **1964,** 5634.
429. Maisey, R. F.: Brit. Pat. 901,892 (1962); *Chem. Abstr.* 1962, **57,** 16397*h*.
430. Codding, D. W., Reid, T. S., Ahlbrecht, A. H., Smith, G. H., Jr., and Husted, D. R.: *J. Polymer Sci.* 1955, **15,** 515.

431. Rozantsev, G. G., Braz, G. I., and Yakubovich, A. Y.: *Zh. Obshchei Khim.* 1964, **34,** 2974; *Chem. Abstr.* 1964, **61,** 15998*a*.
432. Clark, R. F., and Simons, J. H.: *J. Am. Chem. Soc.* 1953, **75,** 6305.
433. Green, M.: *Chem. Ind.* (*London*) **1961,** 435.
434. Bourne, E. J., Tatlow, C. E. M., and Tatlow, J. C.: *J. Chem. Soc.* **1950,** 1367.
435. Sheppard, W. A., and Muetterties, E. L.: *J. Org. Chem.* 1960, **25,** 180.
436. Patton, R. H., and Simons, J. H.: *J. Am. Chem. Soc.* 1955, **77,** 2016, 2017.
437. Henne, A. L., and Stewart, J. J.: *J. Am. Chem. Soc.* 1955, **77,** 1901.
438. Chambers, W. J., and Coffman, D. D.: *J. Org. Chem.* 1961, **26,** 4410.
439. Weygand, F., and Geiger, R.: *Chem. Ber.* 1956, **89,** 647.
440. Schallenberg, E. D., and Calvin, M.: *J. Am. Chem. Soc.* 1955, **77,** 2779.
441. Gilman, H., and Jones, R. G.: *J. Am. Chem. Soc.* 1943, **65,** 1458.
442. Vorozhtsov, N. N., Jr., Barkhash, V. A., Prudchenko, A. T., and Shchegoleva, G. S.: *Zh. Obshchei Khim.* 1965, **35,** 1501; *Chem. Abstr.* 1965, **63,** 14742*f*.
443. Vorozhtsov, N. N., Jr., Barkhash, V. A., Prudchenko, A. T., and Khomenko, T. I.: *Doklady Akad. Nauk SSSR* 1965, **164,** 1046; *Chem. Abstr.* 1966, **64,** 2045*b*.
444. Bergmann, F., and Kalmus, A.: *J. Am. Chem. Soc.* 1954, **76,** 4137.
445. McBee, E. T., Pierce, O. R., and Meyer, D. C.: *J. Am. Chem. Soc.* 1955, **77,** 917.
446. Tarrant, P., and Warner, D. A.: *J. Am. Chem. Soc.* 1954, **76,** 1624.
447. Park, J. D., and Fontanelli, R.: *J. Org. Chem.* 1963, **28,** 258.
448. Harper, R. J., Jr., Soloski, E. J., Tamborski, C.: *J. Org. Chem.* 1964, **29,** 2385.
449. McBee, E. T., Kelley, A. E., and Rapkin, E.: *J. Am. Chem. Soc.* 1950, **72,** 5071.
450. Holbrook, G. W., and Pierce, O. R.: *J. Org. Chem.* 1961, **26,** 1037.
451. Pattison, F. L. M., and Howell, W. C.: *J. Org. Chem.* 1956, **21,** 879.
452. Gilman, H., Tolman, L., Yeoman, F., Woods, L. A., Shirley, D. A., and Avakian, S.: *J. Am. Chem. Soc.* 1946, **68,** 426.
453. Henne, A. L., and Francis, W. C.: *J. Am. Chem. Soc.* 1951, **73,** 3518; 1953, **75,** 992.
454. Haszeldine, R. N.: *J. Chem. Soc.* **1952,** 3423.
455. Pierce, O. R., Meiners, A. F., and McBee, E. T.: *J. Am. Chem. Soc.* 1953, **75,** 2516.
456. Knunyants, I. L., Sterlin, R. N., Yatsenko, R. D., and Pinkina, L. N.: *Izv. Akad. Nauk SSSR* **1958,** 1345; *Chem. Abstr.* 1959, **53,** 6987*g*.
457. Sterlin, R. N., Pinkina, L. N., Knunyants, I. L., and Nezgovorov, L. F.: *Khim. Nauka i Prom.* 1959, **4,** 809, *Chem. Abstr.* 1960, **54,** 10837*i*.
458. Drakesmith, F. G., Richardson, R. D., Stewart, O. J., and Tarrant, P.: *J. Org. Chem.* 1968, **33,** 286.
458a. Henne, A. L., and Nager, N.: *J. Am. Chem. Soc.* 1952, **74,** 650.
459. Nield, E., Stephens, R., and Tatlow, J. C.: *J. Chem. Soc.* **1959,** 166.
460. Vorozhtsov, N. N., Jr., Barkhash, V. A., Ivanova, N. G., Anichkina, S. A., and Andreevskaya, O. I.: *Doklady Akad. Nauk. SSSR* 1964, **159,** 125; *Chem. Abstr.* 1965, **62,** 4045*a*.
461. Harper, R. J., Jr., Soloski, E. J., and Tamborski, C.: *J. Org. Chem.* 1964, **29,** 2385.
462. Pummer, W. J., and Wall, L. A.: U.S. Pat. 3,046,313 (1962); *Chem. Abstr.* 1962, **57,** 15003*a*; 3,150,163 (1964); *Chem. Abstr.* 1964, **61,** 16010*d*.
463. Birchall, J. M., Bowden, F. L., Haszeldine, R. N., and Lever, A. B. P.: *J. Chem. Soc. A*, **1967,** 747.
464. Chambers, R. D., Hutchinson, J., and Musgrave, W. K. R.: *J. Chem. Soc.* **1965,** 5040.
465. Bluhm, H. F., Donn, H. V., and Zook, H. D.: *J. Am. Chem. Soc.* 1955, **77,** 4406.
466. Tarrant, P., Savory, J., and Iglehart, E. S.: *J. Org. Chem.* 1964, **29,** 2009.
467. Nagarajan, K., Caserio, M. C., and Roberts, J. D.: *J. Am. Chem. Soc.* 1964, **86,** 449.

468. Birchall, J. M., Clarke, T., and Haszeldine, R. N.: *J. Chem. Soc.* **1962,** 4977.
469. Coe, P. L., Tatlow, J. C., and Terrell, R. C.: *J. Chem. Soc. C,* **1967,** 2626.
470. Brooke, G. M., and Quasem, M. A.: *J. Chem. Soc. C,* **1967,** 865.
471. Brooke, G. M., Furniss, B. S., Musgrave, W. K. R., and Quasem, M. A.: *Tetrahedron Lett.* **1965,** 2991.
472. Drakesmith, F. G., Richardson, R. D., Stewart, O. J., and Tarrant, P.: *J. Org. Chem.* 1968, **33,** 286.
473. Gilman, H., and Soddy, T. S.: *J. Org. Chem.* 1957, **22,** 1715.
474. Gilman, H., and Gorsich, R. D.: *J. Am. Chem. Soc.* 1956, **78,** 2217; 1957, **79,** 2625.
475. Pierce, O. R., McBee, E. T., and Judd, G. F.: *J. Am. Chem. Soc.* 1954, **76,** 474.
476. Drakesmith, F. G., Stewart, O. J., and Tarrant, P.: *J. Org. Chem.* 1968, **33,** 280.
477. Coe, P. L., Stephens, R., and Tatlow, J. C.: *J. Chem. Soc.* **1962,** 3227.
478. Tamborski, C., and Soloski, E. J.: *J. Org. Chem.* 1966, **31,** 743.
479. Haszeldine, R. N., and Walaschewski, E. G.: *J. Chem. Soc.* **1953,** 3607.
480. Bergmann, E. D., Cohen, S., Hoffman, E., and Rand-Meir, Z.: *J. Chem. Soc.* **1961,** 3452.
481. Connett, J. E., Davies, A. G., Deacon, G. B., and Green, J. H. S.: *J. Chem. Soc. C,* **1966,** 106.
482. Burdon, J., Coe, P. L., and Fulton, M.: *J. Chem. Soc.* **1965,** 2094.
483. McLoughlin, V. C. R., Thrower, J., and White, J. M.: *Chem. Eng. News* 1967, **45**(34), 40.
484. Sterlin, R. N., Yatsenko, R. D., Pinkina, L. N., Knunyants, I. L.: *Izv. Akad. Nauk SSSR* **1960,** 1991; *Chem. Abstr.* 1961, **55,** 13296*h*.
485. Bennett, F. W., Brandt, G. R. A., Emeleus, H. J., and Haszeldine, R. N.: *Nature* 1950, **166,** 225.
486. Cook, D. J., Pierce, O. R., and McBee, E. T.: *J. Am. Chem. Soc.* 1954, **76,** 83.
487. Chen Tsin-Yun, Gambaryan, N. P., and Knunyants, I. L.: *Doklady Akad. Nauk SSSR* 1960, **133,** 1113; *Chem. Abstr.* 1960, **54,** 24385*g*.
488. Darrall, R. A., Smith, F., Stacey, M., and Tatlow, J. C.: *J. Chem. Soc.* **1951,** 2329.
489. McBee, E. T., Campbell, D. H., Kennedy, R. J., and Roberts, C. W.: *J. Am. Chem. Soc.* 1956, **78,** 4597.
490. Dull, D. L., Baxter, I., and Mosher, H. S.: *J. Org. Chem.* 1967, **32,** 1622.
491. Herkes, F. E., and Burton, D. J.: *J. Org. Chem.* 1967, **32,** 1311.
492. Filler, R., and Heffern, E. W.: *J. Org. Chem.* 1967, **32,** 3249.
493. Bergmann, E. D., and Shahak, I.: *J. Chem. Soc.* **1961,** 4033.
494. Gault, H., Rouge, D., and Gordon, E.: *Compt. rend.* 1960, **250,** 1073.
495. Dyachenko, I. A., Livshits, B. R., Gambaryan, N. P., Komarov, V. A., and Abdulganieva, K. A.: *Zh. Vses. Khim. Obshchest.* 1966, **11,** 590; *Chem. Abstr.* 1967, **66,** 7080.
496. Middleton, W. J.: *J. Org. Chem.* 1965, **30,** 1402.
497. Shishkin, G. V., and Mamaev, V. P.: *Zh. Obshch. Khim.* 1966, **36,** 660; *Chem. Abstr.* 1966, **65,** 9010*h*.
498. Henne, A. L., Newman, M. S., Quill, L. L., and Staniforth, R. A.: *J. Am. Chem. Soc.* 1947, **69,** 1819.
499. Reid, J. C., Calvin, M.: *J. Am. Chem. Soc.* 1950, **72,** 2948.
500. Saunders, B. C., and Stacey, G. J.: *J. Chem. Soc.* **1948,** 1773.
501. Blank, I., Mager, J., and Bergmann, E. D.: *J. Chem. Soc.* **1955,** 2190.
502. Bergmann, E. D., and Cohen, S.: *J. Chem. Soc.* **1961,** 4669.
503. Duschinsky, R., Pleven, E., and Heidelberger, C.: *J. Am. Chem. Soc.* 1957, **79,** 4559.
504. Burdon, J., and McLoughlin, V. C. R.: *Tetrahedron* 1964, **20,** 2163.
505. McBee, E. T., Pierce, O. R., Kilbourne, H. W., and Wilson, E. R.: *J. Am. Chem. Soc.* 1953, **75,** 3152.

506. Buchanan, R. L., Dean, F. H., and Pattison, F. L. M.: *Can. J. Chem.* 1962, **40,** 1571.
507. Englund, B.: *Org. Syn.* 1954, **34,** 16, 49; *Coll. Vol.* 1963, **4,** 184, 423.
508. Park, J. D., Vail, D. K., Lea, K. R., and Lacher, J. R.: *J. Am. Chem. Soc.* 1948, **70,** 1550.
509. Tarrant, P., and Brown, H. C.: *J. Am. Chem. Soc.* 1951, **73,** 1781.
510. Koshar, R. J., Simmons, T. C., and Hoffmann, F. W.: *J. Am. Chem. Soc.* 1957, **79,** 1741.
511. Rapp, K. E., Pruett, R. L., Barr, J. T., Bahner, C. T., Gibson J. D., and Lafferty, R. H., Jr.: *J. Am. Chem. Soc.* 1950, **72,** 3642.
512. Knunyants, I. L., German, L. S., and Dyatkin, B. L.: *Izv. Akad. Nauk SSSR* **1956,** 1353; *Chem. Abstr.* 1957, **51,** 8037*f*.
513. Pruett, R. L., Barr, J. T., Rapp, K. E., Bahner, C. T., Gibson, J. D. and Lafferty, R. H., Jr.: *J. Am. Chem. Soc.* 1950, **72,** 3646.
514. Durrell, W., Lovelace, A. M., and Adamczak, R. L.: *J. Org. Chem.* 1960, **25,** 1661.
515. Haszeldine, R. N., and Steele, B. R.: *J. Chem. Soc.* **1957,** 2800.
516. Haszeldine, R. N., and Osborne, J. E.: *J. Chem. Soc.* **1956,** 61.
517. Park, J. D., Seffl, R. J., and Lacher, J. R.: *J. Am. Chem. Soc.* 1956, **78,** 59.
518. Haszeldine, R. N.: *J. Chem. Soc.* **1953,** 922.
519. Tarrant, P., Brey, M. L., and Gray, B. E.: *J. Am. Chem. Soc.* 1958, **80,** 1711.
520. Coffman, D. D., Cramer, R., and Rigby, G. W.: *J. Am. Chem. Soc.* 1949, **71,** 979.
521. Haszeldine, R. N.: *J. Chem. Soc.* **1949,** 2856; **1953,** 3761.
522. Haszeldine, R. N.: *J. Chem. Soc.* **1951,** 2495.
523. Haszeldine, R. N.: *J. Chem. Soc.* **1953,** 3565.
524. Haszeldine, R. N., and Leedham, K.: *J. Chem. Soc.* **1952,** 3483.
525. Haszeldine, R. N.: *J. Chem. Soc.* **1953,** 3359.
526. Haszeldine, R. N., Leedham, K., and Steele, B. R.: *J. Chem. Soc.* **1954,** 2040.
527. Haszeldine, R. N., and Steele, B. R.: *J. Chem. Soc.* **1953,** 1952.
528. Haszeldine, R. N., and Steele, B. R.: *J. Chem. Soc.* **1954,** 923.
529. Haszeldine, R. N., and Steele, B. R.: *J. Chem. Soc.* **1955,** 3005.
530. Haszeldine, R. N., and Steele, B. R.: *J. Chem. Soc.* **1957,** 2193.
531. Haszeldine, R. N., and Steele, B. R.: *J. Chem. Soc.* **1957,** 2800.
532. Henne, A. L., and Kraus, D. W.: *J. Am. Chem. Soc.* 1951, **73,** 1791; 1954, **76,** 1175.
533. Henne, A. L., and Nager, M.: *J. Am. Chem. Soc.* 1951, **73,** 5527.
534. Tarrant, P., and Gillman, E. G.: *J. Am. Chem. Soc.* 1954, **76,** 5423.
535. Tarrant, P., and Lilyquist, M. R.: *J. Am. Chem. Soc.* 1955, **77,** 3640.
536. Tarrant, P., and Lovelace, A. M.: *J. Am. Chem. Soc.* 1954, **76,** 3466.
537. Tarrant, P., and Lovelace, A. M.: *J. Am. Chem. Soc.* 1955, **77,** 768.
538. Tarrant, P., Lovelace, A. M., and Lilyquist, M. R. :*J. Am. Chem. Soc.* 1955, **77,** 2783.
539. Haszeldine, R. N.: *J. Chem. Soc.* **1950,** 2789.
540. Haszeldine, R. N.: *J. Chem. Soc.* **1950,** 3037.
541. Haszeldine, R. N.: *J. Chem. Soc.* **1953,** 922.
542. Haszeldine, R. N.: *J. Chem. Soc.* **1952,** 3490.
543. Haszeldine, R. N., and Leedham, K.: *J. Chem. Soc.* **1952,** 3483.
544. Haszeldine, R. N., and Leedham, K.: *J. Chem. Soc.* **1953,** 1548.
545. Haszeldine, R. N., and Leedham, K.: *J. Chem. Soc.* **1954,** 1261.
546. Leedham, K., and Haszeldine, R. N.: *J. Chem. Soc.* **1954,** 1634.
547. Muramatsu, H., Inukai, K., and Ueda, T.: *J. Org. Chem.* 1965, **30,** 2546.
548. Muramatsu, H., Moriguchi, S., and Inukai, K.: *J. Org. Chem.* 1966, **31,** 1306.
549. Moore, R. A., and Levine, R.: *J. Org. Chem.* 1964, **29,** 1883.

550. Seyferth, D., Mui, J. Y.-P., Gordon, M. E., and Burlitch, J. M.: *J. Am Chem. Soc.* 1965, **87,** 681.
551. Mitsch, R. A.: *J. Am. Chem. Soc.* 1965, **87,** 758.
552. Bartlett, P. D., Montgomery, L. K., and Seidel, B.: *J. Am. Chem. Soc.* 1964, **86,** 616.
553. Solomon, W. C., and Dee, L. A.: *J. Org. Chem.* 1964, **29,** 2790.
554. Karle, I. L., Karle, J., Owen, T. B., Broge, R. W., Fox, A. H. and Hoard, J. L.: *J. Am. Chem. Soc.* 1964, **86,** 2523.
555. Sharts, C. M. and Roberts, J. D.: *J. Am. Chem. Soc.* 1961, **83,** 871.
556. Coffman, D. D., Barrick, P. L., Cramer, R. D., and Raasch, M. S.: *J. Am. Chem. Soc.* 1949, **71,** 490.
557. Barr, D. A., and Haszeldine, R. N.: *J. Chem. Soc.* **1956,** 3416.
558. Misani, F., Speers, L., and Lyon, A. M.: *J. Am. Chem. Soc.* 1956, **78,** 2801.
559. McBee, E. T., Smith, D. K., and Ungnade, H. E.: *J. Am. Chem. Soc.* 1955, **77,** 387.
560. McBee, E. T., Hsu, C. G., Pierce, O. R., and Roberts, C. W.: *J. Am. Chem. Soc.* 1955, **77,** 915.
561. Shozda, R. J., and Putnam, R. E.: *J. Org. Chem.* 1962, **27,** 1557.
562. Anderson, L. P., Feast, W. J., and Musgrave, W. K. R.: *J. Chem. Soc. C,* **1969,** 211.
563. Liu, R. S. H.: *J. Am. Chem. Soc.* 1968, **90,** 215.
564. Liu, R. S. H., and Krespan, C. G.: *J. Org. Chem.* 1969, **34,** 1271.
565. Callander, D. D., Coe, P. L., and Tatlow, J. C.: *Chem. Commun.* **1966** (5), 143.
566. Anderson, L. P., Feast, W. J., and Musgrave, W. K. R.: *Chem. Commun.* **1968,** 1433.
567. Yakubovich, A. Y., Rozenshtein, S. M., Vasyukov, S. E., Tetelbaum, B. I., and Yakutin, V. I.: *Zh. Obshchei Khim.* 1966, **36,** 728; *Chem. Abstr.* 1966, **65,** 8901*b*.
568. Krespan, C. G.: *J. Am. Chem. Soc.* 1961, **83,** 3432.
569. Harris, J. F., Jr., Harder, R. G., and Sausen, G. N.: *J. Org. Chem.* 1960, **23,** 633.
570. Brown, H. C.: *J. Org. Chem.* 1957, **22,** 1256.
571. Boston, J. L., Sharp, D. W. A., and Wilkinson, G.: *J. Chem. Soc.* **1962,** 3488.
572. Belmore, E. A., Ewalt, W. M., and Wojcik, B. H.: *Ind. Eng. Chem.* 1947, **39,** 338.
573. Henne, A. L., and Finnegan, W. G.: *J. Am. Chem. Soc.* 1949, **71,** 298.
574. Gilman, H., and Jones, R. G.: *J. Am. Chem. Soc.* 1943, **65,** 2037.
575. Gething, B., Patrick, C. R., Stacey, M., and Tatlow, J. C.: *Nature* 1959, **183,** 588.
576. Fainberg, A. H., and Miller, W. T., Jr.: *J. Am. Chem. Soc.* 1957, **79,** 4170.
577. Park, J. D., Benning, A. F., Downing, F. B., Laucius, J. F., and McHarness, R. C.: *Ind. Eng. Chem.* 1947, **39,** 354.
578. Prober, M.: *J. Am. Chem. Soc.* 1953, **75,** 968.
579. Cheburkov, Y. A., and Knunyants, I. L.: *Izv. Akad. Nauk SSSR* **1963,** 1573; *Chem. Abstr.* 1963, **59,** 15175*f*.
580. Knunyants, I. L., Cheburkov, Y. A., and Bargamova, M. D.: *Izv. Akad. Nauk SSSR* **1963,** 1393; *Chem. Abstr.* 1963, **59,** 17175*b*.
581. England, D. C., and Krespan, C. G.: *J. Am. Chem. Soc.* 1965, **87,** 4019.
582. Walker, F. H., and Pavlath, A. E.: *J. Org. Chem.* 1965, **30,** 3284.
583. Nield, E., Stephens, R., and Tatlow, J. C.: *J. Chem. Soc.* 1959, 166.
584. LaZerte, J. D., Hals, L. J., Reid, T. S., and Smith, G. H.: *J. Am. Chem. Soc.* 1953, **75,** 4525.
585. Brice, T. J., LaZerte, J. D., Hals, L. J., and Pearlson, W. H.: *J. Am. Chem. Soc.* 1953, **75,** 2698.
586. England, D. C., and Krespan, C. C.: *J. Am. Chem. Soc.* 1966, **88,** 5582.
587. Scherer, O., and Kühn, H.: Ger. Pat. 1,041,937 (1957); *Chem. Zentr.* **1959,** 7886.
588. Madai, H., and Müller, R.: *J. prakt. Chem.* 1963, **19,** 83.
589. Chambers, R. D., Corbally, R. P., Jackson, J. A., and Musgrave, W. K. R.: *Chem. Commun.* **1969,** 127.
590. Camaggi, G., Gozzo, F., and Cevidalli, G.: *Chem. Commun.* **1966,** 313.

591. Haller, J.: *J. Am. Chem. Soc.* 1966, **88,** 2070.
592. Lemal, D. M., Staros, J. V., and Austel, V.: *J. Am. Chem. Soc.* 1969, **91,** 3373.
593. Brown, F., and Musgrave, W. K. R.: *J. Chem. Soc.* **1953,** 2087.
594. Park, J. D., Larsen, E. R., Holler, H. W., and Lacher, J. R.: *J. Org. Chem.* 1958, **23,** 1166.
595. Barr, D. A., and Haszeldine, R. N.: *J. Chem. Soc.* **1956,** 3428.
596. Henne, A. L., and Stewart, J. J.: *J. Am. Chem. Soc.* 1955, **77,** 1901.
597. Husted, D. R., and Kohlhase, W. L.: *J. Am. Chem. Soc.* 1954, **76,** 5141.
598. Barr, D. A., and Haszeldine, R. N.: *J. Chem. Soc.* **1957,** 30.
599. Lewis, E. E., and Naylor, M. A.: *J. Am. Chem. Soc.* 1947, **69,** 1968.
600. Pearlson, W. H., and Hals, L. J.: U.S. Pat. 2,617,836 (1952); *Chem. Abstr.* 1953, **47,** 8770*e*.
601. Young, J. A., and Reed, T. M.: *J. Org. Chem.* 1967, **32,** 1682.
602. Barr, D. A., and Haszeldine, R. N.: *J. Chem. Soc.* **1956,** 3416.
603. Reed, R., Jr.: *J. Am. Chem. Soc.* 1955, **77,** 3403.
604. Tedder, J. M.: *Chem. Rev.* 1955, **55,** 787.
605. Brooke, G. M., Burdon, J., Tatlow, J. C. (and Richardson, R. D.): *J. Chem. Soc.* **1961,** 802.
606. Emmons, W. D., and Pagano, H. S.: *J. Am. Chem. Soc.* 1955, **77,** 89.
607. Emmons, W. D., Pagano, H. S., and Freeman J. P.: *J. Am. Chem. Soc.* 1954, **76,** 3472.
608. Emmons, W. D., and Lucas, G. B.: *J. Am. Chem. Soc.* 1955, **77,** 2287.
609. Sager, W. F., and Duckworth, A.: *J. Am. Chem. Soc.* 1955, **77,** 188.
610. Hawthorne, M. F., Emmons, W. D., and McCallum, K. S.: *J. Am. Chem. Soc.* 1958, **80,** 6393.
*611. Boden, N., Emsley, J. W., Feeny, J., and Sutcliffe, L. H.: *Mol. Phys.* 1964, **8,** 133.
612. Boden, N., Emsley, J. W., Feeny, J., and Sutcliffe, L. H.: *Mol. Phys.* 1964, **8,** 467.
613. Lawrenson, I. J.: *J. Chem. Soc.* **1965,** 1117.
614. Bruce, M. L.: *J. Chem. Soc. A*, **1968,** 1459.
615. Frankiss, S. G.: *J. Phys. Chem.* 1963, **67,** 752.
616. Meyer, L. H., and Gutowski, H. S.: *J. Phys. Chem.* 1953, **57,** 481.
617. Ellenan, D. D., Brown, L. C., and Williams, D.: *J. Mol. Spectry* 1961, **7,** 307.
618. Lee, J. and Sutcliffe, L. H.: *Trans. Faraday Soc.* 1959, **55,** 880.

* References 611–618 will be found on pp. 48–49.

# Author Index

The numbers in the author index are numbers under which the authors are listed on pages 154–169. Since the numbering of the references is consecutive throughout the book (with a very few exceptions), no difficulties should be encountered in locating the authors in the text of the book.


Abrahamsson, S. 171
Adamczak, R. L. 514
Adams, J. H. 405
Aepli, O. T. 12
Ahlbrecht, A. H. 240, 371, 430
Akashi, C. 146
Alderson, T. 312, 372
Allen, D. R. 2, 302
Allison, J. A. C. 33
Anderson, L. P. 562
Anderson, L. R. 80, 566
Andreades, R. A. 248
Andreevskaya, O. I. 460
Anello, L. G. 387
Anichkina, S. A. 369, 460
Antonucci, J. M. 145, 418
Aranda, G. 135
Attaway, J. 95
Austel, V. 592
Avakian, S. 452
Avonda, F. P. 50
Ayer, D. E. 31

Bachman, G. B. 239
Bacon, J. C. 122, 375
Bahner, C. T. 277, 408, 511, 513
Bajzer, W. X. 379
Balz, G. 159
Banks, R. E. 292, 422
Barber, E. J. 7, 83
Barbour, A. K. 8, 231
Bargamova, M. D. 580
Barkhash, V. A. 369, 442, 443, 460
Bartlett, P. O. 552
Barlow, G. B. 8, 87, 337
Barr, D. A. 47, 557, 602
Barr, J. T. 277, 408, 511, 513
Barrick, P. L. 325, 556
Bass, E. A. 201
Baum, M. 253
Baum, M. E. 409
Baxter, I. 490
Beachem, M. I. 380
Beaty, R. D. 164
Becerra, R. 75
Beckert, W. F. 364
Belcher, R. 186
Belmore, E. A. 235, 572
Benkeser, R. A. 334
Bennett, F. W. 485
Benning, A. F. 234, 577
Benoiton, L. 374
Berger, S. 151
Bergman E. 398
Bergmann, E. D. 14, 123, 137, 157, 415, 423, 480, 501, 502
Bergmann, F. 366, 444
Bevan, C. W. L. 417
Bigelow, L. A. 6, 58, 64, 67
Birchall, J. M. 382, 421, 463, 468
Birum, G. H. 107
Bishop, B. C. 67
Bissell, E. R. 53
Blank, I. 123, 501

Bluhm, H. F. 465
Bockemüller, W. 32
Boden, N. 611, 612
Bohler, F. 297
Bollinger, J. M. 74
Booth, H. S. 22
Borden, M. R. 55, 280
Bornstein, J. 54, 55, 280
Borzova, I. 310
Boschan, R. 349
Boston, J. L. 571
Bourne, E. J. 434
Bovey, F. A. 173
Bowden, F. L. 463
Bowers, A. 75
Bradley, C. W. 122, 375
Braendlin, H. P. 252
Braid, M. 78, 329
Brandt, G. R. A. 485
Braz, G. I. 431
Brey, M. L. 519
Brey, W. L. 175
Brice, T. J. 354, 585
Broge, R. W. 554
Bronnert, D. L. E. 401
Brooke, G. M. 293, 294, 470, 471, 605
Brown, E. D. 241
Brown, F. 593
Brown, G. M. 321
Brown, H. C. 276, 390, 402, 509, 570
Brown, J. H. 98
Brown, J. K. 167
Brown, L. C. 617
Brown, R. A. 249
Bruce, M. I. 614
Bryce, H. G. 89, 225
Buchanan, R. L. 506
Buckle, F. J. 15
Buckley, G. D. 48
Burdon, J. 293, 313, 425, 482, 504, 605
Burford, W. B. III 86
Burger, L. L. 7, 83
Burlitch, J. M. 550
Burnard, R. J. 43
Burton, D. J. 491
Buxton, M. W. 391

Cady, G. H. 7, 33, 70, 71, 83
Callander, D. D. 565
Calvin, M. 440, 499
Camaggi, G. 590
Campbell, D. H. 489
Campbell, R. H. 181
Carpenter, C. P. 222
Carpenter, W. 57, 281
Caserio, M. C. 467
Caspar, E. 17
Cassaday, J. T. 12, 375
Cech, M. 275
Cevidalli, G. 590
Chambers, R. D. 77, 82, 128, 259, 289, 294, 328, 383, 428, 464, 589
Chambers, W. J. 60, 438
Cheburkov, Y. A. 579, 580
Chen, P. S. 146
Chen Tsin-Yun 264
Christie, K. O. 143
Clark, J. W. 307
Clark, R. F. 432
Clarke, T. 468
Clayton, J. W. Jr. 4, 223
Clifford, A. F. 72
Codding, D. W. 240, 371, 430
Coe, P. L. 351, 469, 477, 482, 565
Coffman, D. D. 24, 25, 30, 60, 81, 260, 325, 438, 520, 556
Cohen, S. 480, 502
Cohen, S. G. 368
Connett, J. E. 481
Cook, D. J. 486
Cooper, K. A. 141
Corbally, R. P. 589
Cott, W. J. 413
Covington, A. K. 193
Cramer, R. 520
Cramer, R. D. 556
Cross, A. D. 151
Cruickshank, P. A. 185
Cuculo, J. A. 64
Cunningham, W. J. 387
Currie, A. C. 51

Dahmlos, J. 133
Damodaran, V. A. 425
Danek, O. 163
Danilov, S. N. 378
Dannley, R. L. 295, 296
Darrall, R. A. 488
D'Atri, B. 320
Daudt, H. W. 99, 227

Davies, A. G. 481
Davis, R. A. 100
Dawson, J. K. 97
Deacon, G. B. 481
Dean, F. H. 76, 506
Dee, L. A. 345
Denot, E. 75
Detoro, F. E. 6
Dick, J. R. 405
Diesslin, A. R. 9, 88
Dittman, A. L. 243
Domijan, J. D. 404
Donn, H. V. 465
Downing, F. B. 234, 577
Drakesmith, F. G. 289, 458, 472, 476
Dubois, R. J. 387
Duckworth, A. 609
Dull, D. L. 490
Durrell, W. 514
Durst, R. A. 194
Duschinsky, R. 503
Dyachenko, I. A. 495
Dyatkin, B. L. 347, 359, 410, 512

Edgell, W. F. 138
Ehrenberger, F. 199
Ehrenfeld, R. L. 243
Ellenan, D. D. 617
Elliott, J. R. 245
El-Shamy, H. K. 72
Elving, P. J. 189
Emeleus, H. J. 29, 62, 69, 72, 215, 485
Emmons, W. D. 606–608, 610
Emsley, J. W. 174, 611, 612
Engelhardt, V. A. 24, 63, 156
England, D. C. 248, 581, 586
Englund, B. 267, 507
Evans, D. E. M. 177
Evans, F. W. 50
Evans, W. J. 324
Ewalt, W. M. 235, 572

Fager, E. W. 282
Fainberg, A. H. 576
Farah, A. 2
Fawcett, F. S. 24, 25, 30, 81, 260
Fear, E. J. P. 298
Fearn, J. F. 418
Feast, W. J. 291, 324, 562, 566
Feeney, J. 174, 611, 612
Ferm, R. L. 162
Field, D. S. 422
Fields, D. B. 53
Filler, R. 211, 492
Finger, G. C. 126, 161
Finnegan, T. F. 404
Finnegan, W. G. 573
Florin, R. E. 287
Fokin, A. V. 347
Fontanelli, R. 237, 447
Forbes, F. J. 213, 257, 356, 419
Fort, R. C. Jr. 114
Fowler, R. D. 86
Fox, A. H. 554
Fox, C. J. 116, 214
Fox, W. B. 80
Francis, W. C. 210, 453
Frank, A. W. 412
Frankiss, S. G. 615
Freedman, M. B. 42
Freeman, J. P. 90, 607
Freidlina, R. K. 106
Fried, J. H. 41, 42, 124, 256
Frost, L. W. 101, 228
Fuller, G. 51, 553
Fulton, M. 482
Fuqua, S. A. 152
Furin, G. G. 426
Furniss, B. S. 471

Gagnon, J. G. 204
Gale, D. M. 319
Gambaryan, N. P. 264, 487, 495
Ganausia, J. 185a
Gault, H. 494
Gavlin, G. 105
Geiger, R. 439
Gelman, N. E. 202
German, L. S. 410, 512
Gershon, H. 92
Gething, B. 575
Ghirardelli, R. G. 247
Gibson, J. D. 277, 408, 511, 513
Giddey, A. 381
Gilbert, E. E. 387
Gillman, E. G. 534
Gilman, H. 344, 355, 441, 452, 473, 474, 574
Glemser, O. 61
Gogol, V. 310

Goldstein H. 381
Goldwhite, H. 41, 956
Gordon, E. 494
Gordon, M. E. 50, 550
Gorsich, R. D. 474
Gould, D. E. 80
Gozzo, F. 590
Gray, B. E. 519
Green, J. H. S. 481
Green, M. 433
Griffis, C. B. 242
Grindahl, G. A. 379
Griswald, P. H. 282
Grosse, A. V. 7, 37, 45, 83, 336
Grunert, C. 140
Guichon, G. 185a
Gutowski, H. S. 616

Haesler, H. 61
Hall, R. J. 221
Haller, J. 591
Hals, L. J. 279, 309, 584, 585, 600
Hamada, M. 362
Hamilton, J. M. Jr. 86, 224, 244
Hanford, W. E. 325
Harder, R. G. 569
Hardy, C. J. 183
Harper, R. J. Jr. 448, 461
Harris, J. F. Jr. 118, 569
Hasek, W. R. 24, 156
Hass, H. B. 101, 228
Haszeldine, R. N. 47, 72, 85, 104, 212, 215, 261, 271, 290, 292, 303, 308, 326, 340, 343, 353, 382, 421, 422, 454, 463, 468, 479, 485, 515, 516, 518, 521–531, 539, 540–546, 557, 595, 598, 602
Hauptschein, M. 78, 249, 329, 336
Haworth, W. N. 318
Hawthorne, M. F. 610
Heffern, E. W. 492
Heidelberger, C. 503
Henbest, H. B. 136
Henne, A. L. 38, 40, 46, 52, 103, 109, 116, 209, 210, 214, 226, 255, 262, 273, 311, 312, 330, 372, 437, 453, 458a, 498, 532, 533, 573, 596
Henry, M. C. 242
Herken, H. 2
Herkes, F. E. 491
Heyes, J. 294
Hill, H. M. 239
Hilton, D. B. 404
Hine, J. 247, 407
Hinkamp, J. B. 38, 330
Hoan, N. 363
Hoard, J. L. 554
Hodge, E. B. 36
Hodge, H. C. 246
Hoeberg, E. J. 122, 375
Hoffman, C. J. 19, 176
Hoffman, E. 480
Hoffmann, F. W. 125, 510
Holbrook, G. W. 241, 450
Holland, D. G. 427
Holler, H. W. 594
Holtz, D. 216
Hopff, H. 39, 316
Horacek, J. 200, 205
Horicky, M. 23
Horton, C. A. 189
Howell, W. C. 406, 413, 451
Hsu, C. G. 560
Hückel, W. 65
Hudlicky, M. 110, 149, 150, 169, 250, 275, 299, 314, 341, 342
Hughes, E. D. 141
Hume, W. G. 1
Hurka, V. R. 306
Hurst, G. L. 62, 69
Husted, D. R. 240, 371, 597
Husted, D. R. F. 430
Hutchinson, J. 128, 383, 428, 464
Hynes, J. B. 58, 67
Hynes, J. L. 175

Ibanez, L. C. 75
Iglehart, E. S. 466
Ikan, R. 157
Illuminati, G. 270
Inman, C. E. 91, 370
Inukai, K. 547, 548
Ivanova, N. G. 460

Jackson, J. A. 82, 259, 589
Jackson, R. W. 93
Jacquignon, P. 363
Jander, J. 353
Jensen, E. V. 18, 113, 131, 142
Johannson, O. K. 241
Johncock, P. 197

Johnson, L. V. 360
Jonas, H. 26, 119
Jones, R. G. 344, 355, 441, 574
Judd, G. F. 475
Jukes, A. E. 351
Jullien, J. 135

Kagan, F. 93, 153
Kahn, E. J. 120
Kakac, B. 169
Kalmus, A. 366, 444
Karle, I. L. 554
Karle, J. 554
Kasper, J. S. 86
Kauck, E. A. 9, 88
Kaup, R. R. 201
Kaye, S. 262
Kelley, A. E. 449
Kellogg, K. B. 70, 71
Kennedy, R. J. 489
Kent, P. W. 56, 155, 190
Khomenko, T. I. 443
Khrlakyan, S. P. 416
Kilbourne, H. W. 505
Kirchner, P. 183
Kline, G. B. 386
Klockow, M. 158
Knox, L. H. 151
Knunyants, I. L. 263, 264, 284, 338, 347, 359, 410, 416, 456, 457, 484, 487, 512, 579, 580
Kobayashi, Y. 146, 147
Kober, E. 117
Koch, H. F. 42
Kohlhase, W. L. 597
Kolb, B. 183
Kolesnikov, G. S. 315
Koltsov, A. I. 378
Komarov, V. A. 347, 495
Kopecky, J. 150
Körbl, J. 188, 205, 275
Korshak, V. V. 315
Korshun, M. O. 202
Koshar, R. J. 510
Kossatz, R. A. 12
Kost, V. N. 106
Krasuskaya, M. P. 284
Kraus, D. 255
Kraus, D. W. 532
Kraus, E. 275
Krespan, C. G. 79, 319, 564, 568, 581, 586
Krüger, E. 399
Kruse, C. W. 126
Kuhn, H. 587
Kuhn, I. 361
Kuhn, S. J. 365
Kuleshova, N. D. 338

Lacher, J. R. 266, 272, 339, 403, 508, 517, 594
Lafferty, R. H. Jr. 277, 408, 511, 513
LaLande, W. A. Jr. 27
Landault, C. 185a
Langer, J. 26, 119
Larsen, E. R. 594
Latham, J. V. 292
Latif, A. 111
Laucius, J. F. 234, 577
Lawrenson, I. J. 613
LaZerte, J. D. 279, 584, 585
Lea, K. R. 508
Lee, J. 618
Leech, H. R. 10
Leedham, K. 326, 524, 526, 543–546
LeFave, G. M. 393
Lejhancova, I. 299
Lemal, D. M. 592
Leonard, M. A. 186
Lessoff, H. 404
Lever, A. B. P. 463
Levine, R. 549
Lewis, C. J. 96
Lewis, F. E. 599
Lilyquist, M. R. 535, 538
Linch, A. L. 1
Lindsey, R. V. Jr. 248
Lingane, J. J. 192
Link, W. J. 161
Linn, C. B. 37, 45
Litant, I. 86
Litt, M. H. 50
Liu, R. S. H. 563, 564
Livshits, B. R. 495
Loncrini, D. F. 253
Lovelace, A. M. 95, 514, 536–538
Lowe, J. U. Jr. 364
Lucas, G. B. 608
Lukin, M. 295
Lyon, A. M. 558

Lysyj, I. 179

Ma, T. S. 195, 196
Machleidt, H. 158
Madai, H. 301, 588
Mager, J. 501
Magerlein, B. J. 93
Maguire, M. H. 400
Maguire, R. G. 105
Maisey, R. F. 429
Majer, J. R. 172
Malatesta, P. 320
Malichenko, B. F. 112
Mamaev, V. P. 497
Marino, G. 270
Martin, D. G. 153
Martin, J. A. 135
Maxwell, A. F. 6
McArthur, R. E. 394
McBee, E. T. 36, 101, 107, 228, 239, 252, 254, 396, 411, 445, 449, 455, 475, 486, 489, 505, 559, 560
McCallum, K. S. 610
McEwan, T. 220
McGinty, R. L. 300
McHarness, R. C. 234, 577
McLafferty, F. W. 171
McLoughlin, V. C. R. 58, 483, 504
Megson, F. H. 380
Meier, R. 297
Meiners, A. F. 455
Meyer, D. D. 445
Meyer, L. H. 616
Middleton, W. J. 319, 496
Midgley, T. Jr. 226
Miller, C. B. 108
Miller, H. C. 1
Miller, J. 258
Miller, W. T. Jr. 41, 42, 43, 51, 124, 243, 256, 282, 576
Millington, J. E. 139, 321, 406
Million, J. G. 182
Misani, F. 558
Mitsch, R. A. 551
Montgomery, L. K. 552
Moore, G. J. 427
Moore, L. O. 307
Moore, R. A. 549
Morgan, K. J. 167
Moriguchi, S. 548
Morinaga, K. 147
Morita, K. 134
Morton, C. J. 411
Mosher, H. S. 490
Mosteller, J. C. 232
Muetterties, E. L. 24, 435
Mui, J. Y.-P. 550
Müller, R. 301, 588
Müller, W. 305
Munter, P. A. 12
Muramatsu, H. 547, 548
Musgrave, W. K. R. 77, 82, 84, 128, 164, 197, 259, 289, 294, 328, 383, 428, 464, 471, 562, 566, 593
Myers, R. L. 245
Myers, T. 18, 131, 142
Mysov, E. I. 284

Nagarajan, K. 467
Nagasawa, S. 362
Nager, M. 103, 458a, 533
Nakanishi, S. 18, 94, 131, 142
Naylor, M. A. 599
Nerdel, F. 66
Nesmeyanov, A. N. 106, 120
Newkirk, A. E. 44, 238
Newman, M. S. 312, 372, 498
Newton, P. R. 179
Nezgovorov, L. F. 457
Nield, E. 335, 357, 459, 583
Nikolaev, N. S. 203
Nilkantiah, P. M. 332
Norman, J. J. 414
Nunes, F. 55
Nyman, F. 303

Oelrichs, P. B. 220
Oelschlager, H. 283
Oesterling, R. E. 91, 370
Olah, G. 121, 268, 331, 361
Olah, G. A. 74, 365
Olson, P. B. 204
Orestova, V. A. 378
Osborne, J. E. 290, 516
Osina, O. I. 352
Otto, K. 206
Owen, T. B. 554

Pacini, H. A. 20, 132
Pagano, H. S. 606, 607

Paleta, O. 358
Pappas, W. S. 182
Park, J. D. 234, 266, 272, 339, 403, 405, 447, 508, 517, 577, 594
Parker, R. E. 274
Parkhurst, R. M. 152
Parts, L. 138
Patrick, C. R. 575
Pattison, F. L. M. 2, 15, 76, 139, 218, 286, 321, 323, 377, 406, 413, 414, 451, 506
Patton, R. H. 436
Paul, R. C. 215
Pavlath, A. 121, 268, 331, 361
Pavlath, A. E. 20, 132, 143, 582
Pearlson, W. H. 309, 585, 600
Pechanec, V. 200
Pennington, W. A. 207
Percival, W. C. 178
Perry, D. R. A. 291
Perry, J. H. 5
Pervova, E. Y. 263
Pesa, J. 206
Peters, R. 219
Peters, R. A. 2, 221
Petras, D. A. 201
Pierce, O. R. 241, 379, 445, 450, 455, 475, 486, 505, 560
Piggott, H. A. 48
Pike, J. E. 93
Pinkina, L. N. 359, 456, 457, 484
Pittman, A. G. 49
Platonov, V. E. 127
Pleven, E. 503
Plueddeman, E. P. 46
Plunkett, R. J. 233
Pollard, F. H. 183
Porter, J. J. 407
Posta, A. 358
Pozzani, U. C. 222
Pratt, R. J. 113
Preibisch, H. J. 140
Prikryl, J. 376
Prober, M. 243, 367, 578
Prudchenko, A. T. 442, 443
Pruett, R. L. 277, 408, 511, 513
Pummer, W. J. 287, 418, 420, 462
Putnam, R. E. 561

Quadriello, D. 151
Quasem, M. A. 470, 471
Quill, L. L. 498
Qwirtsman, J. 196

Raasch, M. S. 325, 556
Raksha, M. A. 148, 388
Ramler, E. O. 265, 333, 397
Rand-Meir, Z. 480
Rapkin, E. 396, 449
Rapp, K. E. 277, 408, 511, 513
Redmond, W. 165
Reed, R. Jr. 603
Reed, T. M. 601
Reid, J. C. 373, 499
Reid, R. S. 279
Reid, T. S. 240, 371, 430, 584
Reinhardt, C. F. 1
Renwick, J. A. A. 92
Richardson, R. D. 213, 257, 356, 419, 458, 472, 605
Riehl, L. 26, 119
Rigamonti, J. 165
Rigby, G. W. 325
Rigby, W. 520
Roberts, C. W. 489, 560
Roberts, J. D. 385, 386, 467, 555
Robson, P. 58, 288, 423, 424
Roe, A. 160, 161
Roedel, G. F. 245
Rosenberg, H. 232
Rotzsche, H. 184
Rouge, D. 494
Rozantsev, G. G. 431
Rozenshtein, S. M. 567
Ruff, O. 59
Ruggli, P. 17
Ruh, R. P. 100, 273
Rutge, A. J. 13
Rutherford, K. G. 165
Rydon, H. N. 374

Saddy, T. S. 473
Sager, W. F. 609
Saunders, B. C. 3, 15, 129, 217, 286, 322, 384, 401, 500
Sausen, G. N. 569
Savory, J. 77, 328, 466
Saylor, J. C. 241
Schallenberg, E. E. 440
Scherer, O. 586

Scheuer, P. J. 368
Schiemann, G. 159
Schleyer, P. R. 114
Schmersahl, P. 283
Scholberg, H. M. 89
Schram, S. R. 236
Schröder, H. 61
Schults, D. W. 345
Schure, R. M. 211
Scribner, R. M. 348
Seel, F. 26, 119
Seffl, R. J. 272, 517
Seidel, B. 552
Severson, R. G. 334
Seyferth, D. 550
Shahak, I. 14, 137, 493
Sharp, D. L. 49
Sharp, D. W. A. 571
Sharrah, M. L. 266, 339
Sharts, C. M. 555
Shaw, G. 400
Shchegoleva, G. S. 442
Sheehan, J. C. 185
Sheldon, Z. D. 7, 83
Shepard, R. A. 404
Sheppard, N. 221
Sheppard, W. A. 28, 68, 118, 154, 435
Sheveleva, N. S. 202
Shirley, D. A. 452
Shishkin, G. V. 497
Shokina, V. V. 338, 416
Shozda, R. J. 561
Shteingarts, V. D. 352, 426
Sianesi, D. 237
Silversmith, E. F. 385
Silverstein, R. M. 152
Simmons, H. E. 251
Simmons, H. E. Jr. 386
Simmons, T. C. 510
Simons, J. H. 96, 265, 333, 394, 397, 432, 436
Skarlos, L. 54
Slesser, O. 236
Smejkal, J. 150
Smith, D. K. 559
Smith, F. 84, 85, 97, 360, 488
Smith, F. A. 246
Smith, G. 279
Smith, G. H. 584
Smith, G. H. Jr. 240, 371, 430
Smith, H. F. Jr. 222
Smith, R. D. 63, 81, 260
Smith, T. A. 424
Smith, W. C. 24, 25, 34, 63, 73, 156
Solomon, W. C. 345
Soloski, E. J. 448, 461, 478
Speers, L. 558
Stacey, G. J. 129, 322, 384, 500
Stacey, M. 87, 97, 213, 257, 288, 318, 337, 360, 391, 419, 423, 488, 575
Staniforth, R. A. 498
Staros, J. V. 592
Starr, L. D. 161
Steele, B. R. 271, 340, 515, 526–531
Stefani, A. P. 411
Stenhagen, E. 171
Stephens, J. C. 288
Stephens, R. 291, 324, 335, 357, 423, 424, 459, 477, 583
Sterlin, R. N. 359, 456, 457, 484
Stewart, J. J. 209, 437, 596
Stewart, O. J. 458, 472, 476
Steyrmark, A. 201
Stockel, R. F. 380
Stoffer, J. O. 51
Stokes, C. S. 336
Stone, I. 187
Storey, R. A. 82, 259
Stothers, J. B. 323, 377
Streitwieser, A. Jr. 216
Stump, E. C. 242
Suckling, C. W. 98
Suschitzky, H. 269
Sutcliffe, L. H. 174, 611, 612, 618
Suzuki, Z. 134
Swarts, F. 285, 392
Sweeney, R. F. 387
Sweet, R. G. 86
Szinai, S. 415

Taborsky, R. G. 295, 296
Tamborski, C. 427, 448, 461, 478
Tannhauser, P. 113
Tarlin, H. J. 55
Tarrant, P. 95, 122, 276, 375, 389, 390, 402, 446, 458, 466, 472, 476, 509, 519, 534–538
Tatlow, C. E. M. 434

Tatlow, J. C. 8, 87, 97, 177, 213, 257, 288, 293, 313, 324, 335, 337, 351, 356, 357, 360, 391, 419, 423, 424, 425, 434, 459, 569, 477, 488, 553, 565, 575, 583, 605
Taylor, C. W. 354
Taylor, N. F. 190
Teach, E. G. 20, 132
Tedder, J. M. 604
Terrell, R. C. 469
Tetelbaum, B. I. 567
Thomas, B. R. J. 97
Thrower, J. 298, 483
Tiers, G. V. D. 278, 346
Tolman, L. 452
Tolman, V. 115
Troitskaya, V. I. 112
Tronov, B. 399
Trott, P. 109, 311
Truce, W. E. 107
Tullock, C. W. 24, 25, 30, 60, 63
Turner, J. J. 411
Twelwes, R. R. 327
Tyczkowski, E. A. 91, 370
Tyuleneva, V. V. 263

Ueda, T. 547
Uhlir, M. 206
Ungnade, H. E. 559

Vail, D. K. 508
Valkanas, G. 39, 316
Vandenberg, G. E. 93
Vanderwerf, C. A. 162
Varma, P. S. 332
Varsanyi, G. 268, 331
Vasyukov, S. E. 567
Vecchio, M. 102, 230
Veitch, J. 298
Velarde, E. 151
Veldhuis, B. 387
Venkataraman, K. S. 332
Veres, K. 115
Vogel, A. I. 16
Vorozhtsov, N. N. Jr. 127, 130, 352, 369, 426, 442, 443, 460

Waalkes, T. P. 52
Wächter, R. 317
Walaschewski, E. G. 479
Walaschewsky, E. 305
Walborsky, H. M. 253, 409
Walker, F. H. 20, 132, 582
Walkes, T. P. 40
Wall, L. A. 145, 287, 418, 420, 462
Ward, P. F. V. 221
Warner, D. A. 446
Wear, R. L. 354
Weber, C. E. 86
Weiblen, D. G. 166
Weidler-Kubanek, A. M. 50
Weinmayr, V. 11, 327
Welch, A. D. 2
Welch, A. J. E. 48
Welch, Z. D. 101, 228
Wessendorf, R. 158
West, T. S. 186
Wetherhold, J. M. 1
Weygand, F. 183, 439
Whalley, W. B. 98
White, J. M. 483
White, R. W. 413
Wickbold, R. 198
Wiechert, K. 21, 140
Wilding, I. G. E. 322
Wiley, D. W. 251
Wilkinson, G. 571
Willard, H. H. 189, 191
Willenberg, W. 59
Willett, J. E. 374
Williams, D. 617
Willson, J. S. 22
Wilson, E. R. 505
Wilson, L. H. 403
Winter, O. B. 191
Wiper, A. 197
Wojcik, B. H. 235, 572
Wolosinski, H. T. 368
Wood, J. F. 29
Wood, K. R. 56, 155
Woods, L. A. 452
Woolf, C. 108
Woolford, R. G. 323, 377
Wrigley, T. I. 136
Würz, A. 350
Wynn, W. K. 92

Yagupolskii, L. M. 112

Yakobson, G. G. 127, 352, 426
Yakubovich, A. Y. 310, 431, 567
Yakutin, V. I. 567
Yarovenko, N. N. 148, 388
Yatsenko, R. D. 456, 484
Yeoman, F. 452
Yeung, H. W. 258
Youker, M. A. 99, 227
Young, D. E. 80
Young, I. M. 292
Young, J. A. 389, 600

Zahalka, J. 341
Zahn, H. 350
Zakharin, L. I. 106
Zappel, A. 144
Zarubinskii, G. M. 378
Zook, H. D. 465

# Subject Index

The subject index contains names of reactions, names of *types* of compounds, inorganic compounds, and organic reagents. Individual organic compounds are not listed and have to be looked up under the corresponding types.


A

*Acacia georginae* 67
*Acetalization* 107, 108
*Acetylenes*
  addition of
    fluorohaloparaffins 137
    fluoro-olefins 139
    hydrogen fluoride, 22, 23
  reactions with organometallics 124, 126
*Acetylenes, fluorinated*
    see *Fluoroacetylenes*
*Acid catalyzed syntheses* 104–106
*Acidity*
  of fluoroacids 66, 67
  of fluoroalcohols 65, 66
  of hydrogen in fluoroparaffins 65, 67
*Acids, fluorinated*
    see *Fluoroacids*
*Acyl chlorides*
  addition to fluoro-olefins 105
  conversion to acyl fluorides 36
  electrochemical fluorination 30
  Friedel-Crafts synthesis 105
*Acyl fluorides*
  electrochemical fluorination 30
  preparation 36
*Acylation*
  electrophilic (Friedel-Crafts) 105, 106
  nucleophilic 118, 119
*Addition, electrophilic* 83, 84
  of acyl and alkyl halides 105
*Addition, electrophilic* (cont.)
  of halogen fluorides 25, 27, 28
  of halogens 99
  of hydrogen fluoride 21–23
  of hydrogen halides 101, 102
*Addition, free-radical type* 134–137
  of fluorine 23, 24
  of alcohols, aldehydes, and ethers 137
  of halo-olefins 135
  of haloparaffins 134–137
*Addition, nucleophilic* 83–85, 133, 134
  of alcohols 83, 85, 87, 133
  of amines and ammonia 84, 134
  of hydrogen fluoride 22, 83
  of mercaptans 83, 133, 134
  of perfluoroacyl fluorides 28, 83
  of perfluoroaromatics 28, 83
  of phenols 83, 112, 133
  of thiophenols 83, 133
*Alcohols*
  addition to fluoro-olefins 133, 137
  conversion to fluorides 39, 40
  esterification by fluoroacids 107
  reaction with fluorohalo compounds 112
*Alcohols, fluorinated*
    see *Fluoroalcohols*
*Aldehydes*
  addition to fluoro-olefins 137
  basic condensation 129–131
  reaction with organometallics 122, 124, 127, 128

*Aldehydes, fluorinated*
see *Fluoroaldehydes*
*Aldol condensation*
of fluoroaldehydes and fluoroketones 129–131
*Alizarinsulfonates*
preparation of lakes 51–53
*Alkali fluorides*
see also *Cesium*, *Potassium*, *Rubidium*, and *Sodium fluoride*
preparation 15
properties 16
reactions with organic halides 38
*Alkanes*
see *Paraffins*
*Alkenes*
see *Olefins*
*Alkyl chloroformates*
conversion to fluoroformates 36, 37
*Alkyl fluorides*
see *Fluoroparaffins*
*Alkyl fluoroformates*
decomposition to fluorides 39, 40
preparation 36, 37
*Alkylation*
electrophilic (Friedel-Crafts reaction) 105, 106
nucleophilic by fluoro compounds 111–115
*Alkynes*
see *Acetylenes*
*Aluminum bromide*
catalyst in rearrangements 145
replacement of fluorine by bromine 102
*Aluminum chloride*
as catalyst
in Friedel-Crafts synthesis 105, 106
in rearrangements 77, 88, 145
cleavage of perfluoroethers 103
replacement of fluorine by chlorine 102, 103
*Amides, fluorinated*
see *Fluoroamides*
*Amines*
addition to fluoro-olefins 134
conversion to fluorides 42, 44
oxidation to nitro compounds 152
reaction with fluorohalo compounds 114
*Amines, fluorinated*
see *Fluoroamines*
*Amino acids*
synthesis 131, 132
trifluoroacetylation 119
*Amino group*
replacement by
fluorine 42–44
halogens 102
*Ammonia*
reaction with
fluoro-olefins 134
fluoroaromatics 117
*Ammonium fluoroborate* 43
*Analysis* of organic fluoro compounds 46–55
*Anhydrides*
electrochemical fluorination 30
*Anti-Markovnikov addition* 84
*Antimony organometallics* 129
*Antimony pentachloride*
catalyst in fluorination 33, 34
*Antimony pentafluoride*
applications 16
availability 9–12
fluorinating agent 34
preparation 15
properties 16
*Antimony trichloride*
catalyst in fluorination 33
*Antimony trifluoride*
applications 16, 32
availability 9–12
fluorinating agent 33–35
preparation 15
properties 16
*Antimony trifluorodichloride* 34, 35
*Apparatus*
for detection of fluorine 52
for fluorination with metal fluorides 8
for electrochemical fluorination 8
*Arbusov rearrangement* 114
*Arndt-Eistert synthesis* 146
*Aromatic compounds, fluorinated*
see *Fluoroaromatics*
*Aromatic hydrocarbons*
fluorination with
cobalt trifluoride 30
fluorine 29
*Arsenic organometallics* 129
*Arsenic trifluoride*
applications 16

*Arsenic trifluoride* (cont.)
availability 9–12
preparation 15
properties 16
replacement of halogen by fluorine 37, 38
*Aryl iodide difluorides* 14, 24
*Arylation*
by fluoro compounds 115–118
*Atomic refraction*
of fluorine 60, 61
*Autocondensation*
of fluoroketones 130
*Azlactone synthesis* 131

B

*Balz-Schiemann reaction* 43
*Basic condensations*
of fluoro compounds 129–132
*Beilstein test* 50
*Benzenesulfonates*
conversion to fluorides 39
*Benzylic halides*
conversion to fluorides 32, 122, 128
*Benzyne* 122, 127
*Benzyne, tetrafluoro*
formation 128
Diels-Alder reaction 141
*Beta-oxidation*
of fluoroacids 68
*Biological properties*
of fluoro compounds 66–68
*Boiling points*
of fluoro compounds 57–59
*Boron trifluoride*
availability 9–12
catalyst in Friedel-Crafts synthesis 202
preparation 15
properties 16
*Boron trifluoride etherate*
applications 16
availability 9–12
catalyst for decomposition of fluoroformates 40
cleavage of epoxides 38, 39
preparation 15
properties 16
*Bromine*
replacement by hydrogen 91, 93
*Bromine trifluoride*
applications 16
availability 9–12
generator of bromine fluoride 27
preparation 15
properties 16
*Burns*
by fluorine, hydrogen fluoride 5, 6

C

*Carbon, activated*
catalyst for fluorination 33
*Carbon–carbon bond*
cleavage by halogens 101
hydrolytic fission 111
rearrangement 145, 146
*Carbon–fluorine bond*
bond energy 86
hydrogenolysis by
catalytic hydrogenation 89, 90
lithium aluminum hydride 90–92
sodium, zinc 93
hydrolytic fission 87, 108–111
*Carbonation*
of Grignard reagents 122–125
of organolithium compounds 127, 128
*Carbonyl chlorofluoride*
preparation 15
properties 16
reaction with hydroxy compounds 40
*Carbonyl fluoride*
preparation 15
properties 16
reaction with hydroxy compounds 40
*Carbonyl group*
by hydrolysis of *gem*-difluorides 109, 110
catalytic hydrogenation 89
reaction with
Grignard reagents 120, 122, 124
organolithium compounds 127
*Carbonyl oxygen*
replacement by fluorine 41
*Carboxyl group*
by hydrolysis of trifluoromethyl group 110
conversion to trifluoromethyl group 41, 42
elimination by decarboxylation 144, 145

*Carboxyl group* (cont.)
reaction with organometallics, 122, 123, 125
reduction with lithium aluminum hydride 90
replacement by halogen 101
*Carboxylic acids, fluorinated*
see *Fluoroacids*
*Catalytic hydrogenation*
of fluoro compounds 89, 90
*Cerium alizarin complexonate* 51
*Cesium fluoride*
applications 16
availability 9–12
preparation 15
properties 16
replacement of halogen by fluorine 37, 38
*Chemical resistance*
of metals and plastics 6, 7
of fluorinated plastics 6, 7, 76
*Chlorine* (*see also Halogens*)
oxidation of isothiocyanates and isothiuronium salts 98
replacement by
fluorine 33–38
hydrogen 91, 92
replacement of fluorine by aluminum chloride 103
*Chlorine trifluoride*
applications, preparation, properties 15, 16
availability 9–12
*Chloro derivatives*
conversion to fluoro derivatives,
see under *Fluoro derivatives*
*Chlorofluoromethanes*
see *Freons*
*Chlorotrifluorotriethylamine*
fluorinating agent 40, 41
*Cholinesterase*
inhibition by fluorophosphates 66
*Chromic acid*
oxidation of fluoro compounds 95, 96
*Chromyl chloride*
oxidation of fluoro compounds 97
*Claisen condensation* 131, 132
*Cleavage*
of carbon chain by chlorine 101, 103
of epoxides 38, 39
*Cleavage* (cont.)
of esters 39
of ethers 39
*Cobalt trifluoride*
applications, preparation, properties 15, 16
availability 9–12
fluorination 30
reaction with nitriles 26
reactor for fluorination 8
*Combustion*
of organic fluorides 51, 54, 55
*Copolymers, fluorinated* 74
*Copper*
coupling of fluorohalo compounds 129, 143
organometallic compounds 129
*Copper nitrate*
oxidizing agent 97
*Corrosion of metals and plastics*
by fluorine and hydrogen fluoride 7, 8
*Corticosteroids, fluorinated 78*
*Curtius degradation* 146
*Cyanohydrin synthesis* 130
*Cycloadditions* 137–142
*Cyclo-olefins, Cycloparaffins*
see *Fluoro-olefins, Fluoroparaffins*

**D**

*Decarboxylation*
of fluorinated acid salts 87, 88
*Decomposition*
of diazonium fluoroborates, fluorophosphates, and fluorosilicates 43
of fluoroformates 40
*Defluorination* 143
*Dehalogenation* 142, 143
*Dehydration*
of fluoroacids 144
of fluoroalcohols 82
*Dehydrobenzene* 122, 127
*Dehydrohalogenation* 143, 144
*Density*
of fluoro compounds 60
of fluoro polymers 75
*Detection*
of fluoride ion 50–52
of fluorine, nitrogen, and sulfur in organic fluoro compounds 51, 52

*Determination*
of fluorine 52–54
of other elements 54, 55
*Diazo compounds*
by oxidation of hydrazones 97
*Diazo group*
replacement by halogens 42
hydrogen halides 42, 102
*Diazomethane*
addition to fluoro-olefins 139
*Diazonium salts*
formation and decomposition 42, 43, 102
*Diazotization*
of fluoroamines 104
*β-Dicarbonyl compounds, β-Dicarboxyl compounds*
reaction with perchloryl fluoride 30
*Dichapetalum cymosum, D. toxicarium*
toxic principle 67
*Dielectric constant*
of fluoro compounds 61
of fluoro polymers 75
*Diels-Alder reaction, Diene synthesis* 139–142
*Difluoro derivatives, gem.*
hydrolysis 109, 110
preparation 41
*2,4-Dinitrofluorobenzene*
preparation 37
reactivity 116
*Diphenyltrifluorophosphorane*
fluorinating agent 40
preparation 25
*Directing effect of fluorine*
in additions 84, 85, 101, 133
in substitutions 85, 100
*Dissociation constants*
of fluoroacids, fluoroalcohols, and fluoroamines 65–67
*Double bond*, reactions at
see *Fluoro-olefins*
*Drugs, fluorinated* 78

E

*Elastomers, fluorinated* 70–77
*Electrochemical fluorination*
apparatus 8
of acyl halides and anhydrides 30
*Electrolysis*
in anhydrous hydrogen fluoride 30
of fluoroacids 98
*Electrophilic additions*
see *Addition, electrophilic*
*Electrophilic substitution* 100, 103, 104, 106
*Elementary analysis*
of fluoro compounds 54, 55
*Elimination*
of carbon dioxide 101, 144, 145
of fluorine 143
of halogens 142, 143
of hydrogen halides 143, 144
of water 145
*Emulsion polymerization 74*
*Enamines*
reaction with perchloryl fluoride 30, 31
*Enol ethers*
reaction with perchloryl fluoride 30, 31
*Epoxides*
by addition of oxygen 95
by reaction with trifluoroperoxyacetic acid 153
cleavage to fluorohydrins 38, 39
*Equipment for fluorination* 6–8
*Esterification*
of fluoro compounds 106, 107
*Esters*
by oxidation of ketones 153
Claisen condensation 131, 132
reaction with Grignard reagents 122, 123
*Esters, fluorinated*
see *Fluoroesters*
*Ethers*
cleavage to fluorides 38
*Ethers, fluorinated*
see *Fluoroethers*
*S-Ethyl thioltrifluoroacetate*
trifluoroacetylating agent 119

F

*Ferric chloride*
fluorination catalyst 33
*Fire extinguishers* 70
*Fluon, Fluorel*, 76
*Fluoride ion*
detection and determination 51–53

*Fluorides, inorganic*
application 16
availability 9–12
preparation 15
properties 16
*Fluorides, organic*
analysis 46–55
nomenclature 18, 19
preparation, survey 44, 45
reactions, survey 148–150
*Fluorinating agents* 9–17
*Fluorinated drugs* 77, 78
*Fluorination*
apparatus 8
by electrochemical process 30
with cobalt trifluoride and silver difluoride 30
with elementary fluorine 29
*Fluorine*
addition to
aromatic compounds 29
olefins 23, 24
other unsaturated systems 24–26
elimination by metals 143
hydrolysis 108–111
replacement by
amino group 114, 117
halogens 102, 103
hydrogen 89–91
by catalytic hydrogenation 89, 90
with lithium aluminum hydride 90, 91
with sodium 93
by hydroxyl 108, 109
substitution for
amino group 42, 43
halogens 31–38
hydrogen 29–31
oxygen 38–42
*Fluorine, elemental*
analysis 50
applications 16, 29
availability 9–12
corrosion by 7
fluorination 29
handling 5
preparation 13
properties 14, 16
*Fluorine–carbon bond*
see *Carbon–fluorine bond*
*Fluorine chemistry*
literature 2–5
symposia 1
*Fluoro-*(see also *Perfluoro-*)
*Fluoroacetals*
preparation 107, 108, 112
*Fluoroacetates*
biological properties 66–68
*Fluoroacetylenes*
addition of
haloparaffins 137
hydrogen fluoride 22, 23
hydrogen halides 102
addition to fluoro-olefins and dienes 140, 141
by alkylation of acetylene 115
by dehalogenation 143
by Grignard synthesis 124
preparation 126, 143
tetramerization 142
*Fluoroacids* (see also *Perfluoroacids*)
by Grignard syntheses 122–125
by hydrolysis of
fluoroethers 110
fluoroparaffins 108
by Knoevenagel synthesis 131
by organometallic syntheses 127, 128
by oxidation with
oxygen 95
nitric acid 99
potassium permanganate 96, 97
sodium dichromate 97
by rearrangement 123, 127
conversion to acyl halides 102
dissociation constants 66, 67
electrolytic coupling 98
esterification 107
reaction with organolithium compounds 125
*Fluoroacids, salts*
conversion to esters 107
decarboxylation 87, 88, 144, 145
*Fluoroacyl azides*
Curtius degradation 146
*Fluoroacyl halides*
acylation 118, 119
Arndt-Eistert synthesis 146
conversion to fluoroacyl azides 146
Friedel-Crafts synthesis 106
preparation 102

*Fluoroalcohols*
acylation 118
by Grignard synthesis 120, 122–125
by reaction with organolithium compounds 126, 127
conversion to halides 102
dehydration 82
dissociation constants 66
esterification 106, 107
rearrangement to fluoroacids 123, 127
*Fluoroaldehydes*
acetalization 107, 108
aldol condensation 129–131
by Grignard synthesis 120, 122–124
by organometallic syntheses 128
by oxidation of fluoroaromatics 97
*Fluoroalkylarsines* 129
*Fluoroalkylphosphines* 129
*Fluoroalkylstibines* 129
*Fluoroamides*
by Curtius degradation 146
by reaction of fluoro-olefins with amines and ammonia 134
by reaction of fluoroesters with ammonia 119
conversion to esters 107
reduction with lithium aluminum hydride 90
*Fluoroamines*
basicity 63, 65
by addition of amines to fluoro-olefins 134
by alkylation with fluorohaloparaffins 114
by arylation 117
by reduction of
amides 90
isocyanates and urethanes 92
nitriles 90
diazotization 104
dissociation constants 65
hydrolysis to fluoroamides 110
oxidation to nitro compounds 153
*Fluoroamino acids*
synthesis 131, 132
*Fluoroaromatics*
arylation by 115–118
Friedel-Crafts synthesis 105
Grignard synthesis 121, 122, 124, 125
halogenation 100
*Fluoroaromatics* (cont.)
hydrolysis 109–111
nitration 103, 104
organolithium compounds 126, 128
oxidation 96, 97
reduction 89, 90
sulfonation 104
*Fluoroazides*
by alkylation 114
Curtius degradation 146
*Fluoroborates* 15, 16
conversion of amines to fluorides 43
*Fluoroboric acid* 15, 16
applications, preparation, properties 15, 16
conversion of amines to fluorides 43
*Fluorobromo-*, *Fluorochloro-*, *Fluoroiodo-*
see *Fluorohalo-*
*Fluorocarbons*
see *Perfluoro compounds*
*Fluoro compounds*
see individual types
*Fluoro copolymers* 74
*Fluorocyclo-olefins*, *Fluorocycloparaffins*
see *Fluoro-olefins*, *Fluoroparaffins*
*Fluorodiazo compounds*
by diazotization of fluoroamines 104
reaction with
hydrogen fluoride 42
hydrogen iodide 102
*Fluorodiazoketones*
reaction with hydrogen fluoride 42
Wolff rearrangement 146
*Fluorodiazonium compounds*
decomposition 43
preparation 43, 104
*Fluorodienes*
see *Fluoro-olefins*
*Fluoro-β-diketones*
by Claisen condensation 131
*Fluoroenamines* 114, 134
*Fluoroenolethers* 109, 112, 133
*Fluoroesters*
by acylation 118
by alkylation 115
by Arndt-Eistert synthesis 146
by esterification 106, 107
by hydrolysis of fluoro ethers 110
by reaction with diazoalkanes 107
Claisen condensation 131, 132

*Fluoroesters* (cont.)
conversion to fluoroamides 119
Grignard synthesis 123, 124
organometallic synthesis 128
Reformatsky synthesis 128
*Fluoroethers*
by addition of alcohols or phenols to fluoro-olefins 109, 112, 133
by alkylation of alcohols or phenols 111, 112
by arylation of alcohols 116
cleavage with aluminum chloride 87, 103
hydrolysis to fluoroacids, fluoroesters 110
*Fluoroform reaction* 84, 111
*Fluoroformates*
by reaction with carbonyl difluoride or chlorofluoride 40
by reaction of chloroformates 36, 37
decomposition to fluorides 39, 40
*Fluorohalo-olefins*
addition to fluoro-olefins 135
addition of hydrogen halides 101, 102
by addition of hydrogen halides 102
reaction with
alcohols 112
amines 114
Grignard reagents 121, 123
mercaptans 114
organolithium compounds 125–127
phenols 112
phosphite 114
*Fluorohaloparaffins*
addition to
fluoro-olefins 105, 135, 136
acetylenes 137
alkylating agents 113–115
dehalogenation 142, 143
dehydrohalogenation 144
Friedel-Crafts reaction 82, 83, 105, 106
isomerization 88, 145
preparation 98–102
reaction with magnesium 122
rearrangement 145
*Fluoroheterocyclics*
preparation 37, 43
reaction with
magnesium 125
nucleophiles 109, 116, 118
reduction 90, 92
*Fluorohydrazines* 117, 118
*Fluoroisocyanates*
by Curtius degradation 146
by Hofmann degradation 147
reduction 92
*Fluoroisothiuronium salts*
oxidation to sulfonylchlorides 98
*Fluoroketenes* 144, 145
*Fluoro β-ketoesters*
Claisen condensation 131, 132
*Fluoroketones*
aldol condensation 129–131
by addition of acyl halides to fluoro-olefins 105
by Friedel-Crafts synthesis 105, 106
by hydrolysis of fluoro-olefins 108
by ketonic fission 132
by oxidation of fluoro-olefins 96
by reaction with organometallics 125
cyanohydrin synthesis 130
Grignard synthesis 120, 122, 123
Knoevenagel condensation 82
reduction 89, 90, 93
Wittig reaction 130
*Fluoromercury compounds* 128
*Fluoronitriles*
basic condensation 131
by reaction of
fluoro-olefins with ammonia 134
fluorohaloparaffins with cyanide 115
conversion to fluoroesters 107
Diels-Alder reaction 141
Grignard synthesis 120
reduction 90, 120
*Fluoronitro compounds*
by aldol condensation 129
by oxidation with trifluoroperoxyacetic acid 152
catalytic hydrogenation and reduction 89
preparation 103, 104
*Fluoronitroso compounds*
addition to fluoro-olefins 141, 142
preparation 104
*Fluoro-olefins*
addition of
alcohols 112, 133, 137
aldehydes 137
amines and ammonia 137
ethers 137

*Fluoro-olefins* (cont.)
addition of (cont.)
fluorohalo-olefins and fluorohalo-paraffins 135–137
halogens 99
hydrogen halides 101, 102
mercaptans and thiophenols 133, 134
nitrogen tetroxide 103
olefins 135, 137
by addition of hydrogen fluoride 22
by decarboxylation 88, 144, 145
by dehalogenation 142, 143
by dehydrohalogenation 143, 144
by Wittig reaction 130
catalytic hydrogenation 89
cycloadditions 137–142
dimerization 138, 139
hydrolysis 108–110
monomers for plastics 73
nucleophilic addition of hydrogen fluoride 22
oxidation 95, 96
pyrolysis 147
reaction with
Grignard reagents 121–123
nucleophiles 109, 112–114
organolithium compounds 125–127
*Fluoro-oxazetidines* 139, 147
*Fluoroparaffins*
by addition of
fluorine 23, 24
hydrogen fluoride 21–23
decarboxylation of perfluoroacids 85, 86
fluorination of chloroparaffins 31–38
Friedel-Crafts synthesis 105, 106
*Fluorophenol ethers* 116
*Fluorophosphates*
applications 78
biological properties 66
conversion of amines to fluorides 43, 44
nomenclature 19
*Fluorophosphines*
nomenclature 19
*Fluorophosphoranes*
fluorinating agents 40
nomenclature 19
preparation 25
reaction with alcohols 40
*Fluoro plastics*, *Fluoro polymers* 70–77
*Fluorosilanes* 32, 33
*Fluorosilicates*
conversion of amines to fluorides 43
fluorinating agents 37, 38
*Fluorosteroids*
by cleavage of epoxides 39
by replacement of
halogen in halosteroids 35
hydrogen by fluorine 31
medicinal applications 78
*Fluorosulfides*
by addition of mercaptans or thiphenols 133, 134
by alkylation 114
by arylation 117
by reaction of sulfur with organomercurials 128
oxidation to sulfoxides and sulfones 97
*Fluorosulfinates of alcohols*
preparation and decomposition to alkyl fluorides 40
*Fluorosulfones*
by oxidation of fluorosulfides 97
*Fluorosulfonic acid* 10–12, 15, 16
*Fluorosulfonyl chlorides* 97, 98
*Fluorosulfoxides* 97
*Fluorothiocyanates*
by acylation 119
by alkylation 113
oxidation to fluorosulfonyl chlorides 97, 98
*Fluorothiophenols*
desulfuration 90, 117
preparation 117, 126
*Fluorouracil* 78
*Fluothane* 77
*Fluoxidin* 78
*Free-radical addition*
to fluoroacetylenes 137
to fluoro-olefins 134–137
*Freons*
applications 69–71
discovery 1
nomenclature 19, 20
preparation 69, 70
properties 72
*Friedel-Crafts synthesis* 105, 106

G

*Gas (-liquid) chromatography* 50

*Gidyea*
toxic principle 67
*Gifblaar*
toxic principle 67
*Glass*
resistance to fluorine and fluoro compounds 6, 7
*Glass etching*
test for fluoride 52
*Glaucoma treatment* 78
*Glycols*
by oxidation with trifluoroperoxyacetic acid 153
*Gold*
catalyst for fluorination 29
*Grignard reagents*
preparation, reactions 120–124
reductive action 93, 120

H

*Halogenation*
of fluoro compounds 98–103
*Halogens*
see also *Bromine*, *Chlorine*, *Iodine*
addition to fluoro-olefins 99
elimination 142, 143
rearrangement in fluorohaloparaffins 88, 145
replacement by fluorine 31–38
using alkali fluorides 32, 36–38
using antimony fluorides 32–35
using arsenic trifluoride 37, 38
using hydrogen fluoride 32, 33
using mercury fluorides 32, 35, 36
using silver fluorides 32, 35
using sodium fluorosilicate 37, 38
using thallous fluoride 37
replacement of
carboxyl 101
diazo group 102
hydrogen 99, 100
lithium 100
*Haloparaffins*
addition to
acetylenes 137
olefins 134–136
*Halothane* 77
*Handling*
of fluorine and hydrogen fluoride 5, 6
*Hexafluorobenzene*
see *Perfluoroaromatics*
*History of fluorine chemistry* 1
*Hofmann degradation* 147
*Hunsdieckers' method* 101
*Hydrates of fluoroketones* 81
*Hydrazine*
reaction with fluoroaromatics 117, 118
*Hydriodic acid*
reduction of fluoro compounds 93
*Hydrofluoric acid*
conversion of alcohols to fluorides 39
corrosion of materials 7
distillation 13
formation and decomposition of diazonium salts 43
*Hydrogen*
determination in organic fluorides 54, 55
replacement by
fluorine 28–31
halogens 99, 100
*Hydrogen bromide, chloride, iodide*
see *Hydrogen halides*
*Hydrogen cyanide*
addition to fluoro ketones 130
*Hydrogen fluoride (anhydrous)*
addition to
acetylenes 22, 23
olefins 21, 22
unsaturated compounds 23
applications 16, 32
availability 9–12
cleavage of epoxides 38
corrosion of materials 7
decomposition of diazonium salts 43
electrolysis of organic compounds 30
elimination by alkalies 144
elimination from
fluoroamines 134
fluoroethers and sulfides 133
handling and injuries 5, 6
preparation 12
properties 16
replacement of halogen by fluorine 32, 33
vapor pressure 17
water content 55
*Hydrogen halides*
addition to
fluoroacetylenes 102

*Hydrogen halides* (cont.)
addition to (cont.)
fluoro-olefins 101, 102
elimination by alkalies 144
replacement of
hydroxyl 102
diazo group 102
*Hydrogen peroxide*
preparation of trifluoroperoxyacetic acid 152
*Hydrogenation, catalytic* 89, 90, 94
*Hydrogenolysis* of carbon–fluorine bond 89, 90
*Hydrolysis* of fluoro compounds 108–111
*Hydroxyl*
replacement by
fluorine 39, 40
halogen 102
*Hyperconjugation* 84
*Hypohalites*
alkaline, Hofmann degradation 147
organic 27, 28

I

*Inductive effect of fluorine* 80–84
*Infrared spectroscopy* 46, 47
*Inorganic fluorides*
availability 9–12
preparation 15
properties and applications 16
*Interhalogen compounds*
addition to fluoro-olefins 25, 27, 28, 99
*Iodine* (see also *Halogens*)
activation of mercurous fluoride 35, 36
replacement by hydrogen with lithium aluminum hydride 91
*Iodine monofluoride*
formation and reactions 27, 28
*Iodine pentafluoride*
applications 16
availability 9–12
preparation 15
properties 16
reactions 25, 27
*Iron*
defluorination 143
reduction 93, 94
resistance to fluorine and hydrogen fluoride 7
*Isomerizations*
of fluorohaloparaffins 88, 145
of fluoro-olefins 145, 147
of perfluoroaromatics 146
of unsaturated fluoroalcohols 123, 126, 127

K

*Karl Fischer method*
for determination of water 55
*Ketones*
basic condensation 129–131
by Friedel-Crafts synthesis 105, 106
by hydrolysis of *gem*-difluorides 109, 110
Claisen condensation 131
Grignard synthesis 120, 123, 125
oxidation to esters with trifluoroperoxy-acetic acid 153
*Ketones, fluorinated*
see *Fluoroketones*
*Knoevenagel reaction* 131
*Kolbe synthesis*
(anodic coupling of fluoroacids) 98
*Krebs cycle*
inhibition by fluoroacetate 67

L

*Lactones*
by oxidation of ketones with trifluoro-peroxy acetic acid 153
*Lanthanum nitrate*
titration of fluorides 54
*Lead dioxide*
reaction with hydrogen fluoride 23, 24
*Lead nitrate*
oxidation of fluoroaromatics 97
*Lead tetraacetate*
reaction with hydrogen fluoride 24
*Lithium aluminum hydride*
reduction of fluoro compounds 90–92, 94
*Lithiumorganic compounds* 125–128

M

*Magnesium*
dehalogenation 143

*Magnesium* (cont.)
formation of Grignard reagents 120–125
*Manganese trifluoride*
applications 16
availability 9–12
fluorination of nitriles 25
preparation 15
properties 16
*Mass spectroscopy* 47, 48
*Mechanisms*
see specific reactions
*Meerwein-Ponndorf reduction* 93
*Melting points*
of fluoro compounds 57
*Mercaptans*
reactions with fluoro compounds 114, 117, 134
*Mercurials, organo* 128
*Mercury fluorides*
addition of fluorine to nitriles 25, 26
applications 16
availability 9–12
preparation 15
properties 16
replacement of halogen by fluorine 35, 36
*Mesomeric effect* 84, 85
*Methanesulfonyl fluoride*
conversion of alcohols to fluorides 39
*Methoxyflurane* 77, 78
*Michael addition* 132
*Migration*
of halogens in fluorohaloparaffins 88, 145
of fluoroalkyl groups in perfluoroaromatics 145
*Migratory aptitudes*
in oxidation of ketones with trifluoroperoxy acetic acid 152, 153
*Mineralization of fluorine* 53, 54
*Molding of fluoroplastics* 74
*Molecular rearrangements* 145–147
*Molecular weight determination* 47
*Monel metal*
resistance of fluorine and hydrogen fluoride 7
*Monographs on fluorine chemistry* 2–5
*Monomers, fluorinated*
polymerization 74
preparation and properties 72, 73

**N**

*Neoglaucit* 78
*Nerve poisons* 66, 78
*Nickel*
catalyst for hydrogenation 90
defluorinating agent 143
resistance to fluorine and hydrogen fluoride 7
*Nitration of fluoro compounds* 103, 104
*Nitric acid*
nitration of fluoro compounds 103, 104
oxidation of fluoroaromatics 97
*Nitric oxide*
reaction with perfluoroalkyl iodides 104
*Nitriles*
addition of fluorine 25, 26
arylation 118
Claisen condensation 131
*Nitriles, fluorinated*
see *Fluoronitriles*
*Nitro compounds*
aldol condensation with fluoroaldehydes 129
by oxidation with trifluoroperoxyacetic acid 152
reactions with perchloryl fluoride 30
*Nitrogen*
determination in organic fluorides 52, 53
replacement by fluorine 42–44
*Nitrogen oxides*
reaction with fluoro compounds 103
*Nitroperfluoro compounds* 103
*Nitrosation*
of fluoro compounds 104
*Nitrosyl chloride*
reaction with fluoro compounds 104
*NMR spectroscopy* 48, 49
*Nomenclature of organic fluoro compounds* 18–20
*Nucleophilic addition to fluoro-olefins* 22, 83, 133, 134

**O**

*Olefins*
addition of
fluorine 24
halogen fluorides 27
haloparaffins 134–136

*Olefins* (cont.)
addition of (cont.)
hydrogen fluoride 21
addition to fluoro-olefins 138, 140
conversion to epoxides and glycols by trifluoroperoxyacetic acid 153
*Oleum*
hydrolysis of fluorohaloparaffins 108
*Organic fluorides*
analysis 46–55
fluorinating agents 14, 17
interconversions 148–150
preparation 21–45
reactions 80–151
*Organocopper compounds* 129
*Organolithium compounds* 125–128
*Organomagnesium compounds* 120–124
*Organomercury compounds* 128
*Organometalloids* 129
*Organophosphorus compounds* 129, 130, 142
*Organozinc compounds* 128
*Orientation by fluorine*
in electrophilic addition 101
in nucleophilic addition 22, 133
in substitution 84, 85
*Oxidation*
by trifluoroperoxyacetic acid 152, 153
of organic fluoro compounds 95–98
*Oxiranes*
see *Epoxides*
*Oxygen*
combustion of organic fluorides 54, 55
oxidation of fluoro compounds 95
replacement by fluorine 38–42
replacement of fluorine in hydrolysis 108–111

P

*Palladium*
catalyst for hydrogenation 89, 90
*Parachor*
of fluoro compounds 62
*Paraffins*
fluorination with
cobalt trifluoride 30
elemental fluorine 29
silver difluoride 30
*Parametasone* 78
*Perchloryl fluoride*
applications 16, 31, 32
availability 9–12
preparation 15
properties 16
reaction with benzene 106
replacement of hydrogen 31, 32
*Perfluoroacetylenes*
cycloadditions 140–142
*Perfluoroacids and their salts*
by electrochemical fluorination 30
decarboxylation 87, 88, 144, 145
properties 64, 66, 67
reaction with
halogens 101
nitrosyl chloride 104
*Perfluoroacyl fluorides*
by electrochemical fluorination 30
reaction with
chlorine fluoride 28
perfluoro-olefins 28
*Perfluoroaldehydes*
by Grignard synthesis 123, 124
synthesis with organolithium compounds 128
Erlenmeyer synthesis 131
*Perfluoro (alkylalkyleneimines)*
by fluorination of nitriles 25, 26
by pyrolysis 147
*Perfluoroalkylarsines* 129
*Perfluoroalkylcarbamates*
by Curtius degration 146
reduction 92
*Perfluoroalkyl halides*
addition to
acetylenes 137
olefins 136
by Hofmann degradation 147
by Hunsdieckers' method 101
by reaction or organozinc compounds 128
properties 57
*Perfluoroalkylisocyanates*
by Curtius degradation 146
by Hofmann degradation 147
reduction with lithium aluminum hydride 92
*Perfluoroalkyllithium compounds* 127
*Perfluoroalkylmagnesium compounds* 122, 123

*Perfluoroalkyloxazetidines*
decomposition 147
formation 139
*Perfluoroalkylphosphines* 129
*Perfluoroalkylstibines* 129
*Perfluoroalkylurethanes*
by Curtius degradation 146
reduction 92
*Perfluoroalkylzinc compounds* 128
*Perfluoroamines* 117
*Perfluoroaromatics*
by defluorination 143
by fluorination of perchloroaromatics 37
Grignard compounds 121, 125
hydrolysis 109
organolithium compounds 126, 128
reaction with
alcohols 116, 117
ammonia 117
hydrazine 117
hydrogen sulfide 117
mercaptans and thiophenols 117
perfluoro-olefins 28
potassium hydroxide 109
sodium alkoxides 116
reduction 90, 92
*Perfluorocarboxylic acids*
see *Perfluoroacids*
*Perfluoro compounds*
hydrolysis 110, 111
nomenclature 19
preparation 29
properties 57*ff*
*Perfluoroethers*
cleavage by aluminum chloride 87, 103
*Perfluorohalo-olefins*
addition to olefins 135–137
by elimination 143
*Perfluoroketenes* 144
*Perfluoroketones*
addition of ammonia, diazomethane, hydrogen fluoride, water 81
basic condensation 129, 131
preparation by
Grignard synthesis 123
oxidation of fluoro-olefins 96
*Perfluoronitriles*
Grignard synthesis 120
reduction 120
*Perfluoro-olefins*
addition reactions 133–142
by decarboxylation 87, 88, 144, 145
by decomposition of fluoro-organometallics 127, 128
cycloadditions 139, 140
Diels-Alder reaction 140, 142
hydrogenation 89
reaction with
alcohols 133
aldehydes 137
aluminum halides 102
amines 114, 134
ammonia 134
lithium aluminum hydride 91
mercaptans 114, 133, 134
organometallics 121
perfluoroacyl fluorides 28
perfluoroaromatics 28
water and alkalies 108
*Perfluoroparaffins*
by fluorination of paraffins or aromatics 29, 30
halogenolysis 101
hydrolysis 110
properties 56, 57, 61
pyrolysis 147
*1H-Perfluoroparaffins*
by decomposition of
perfluoroacid salts 87, 88, 144, 145
organometallics 124, 127, 128
by Hofmann degradation 147
*Perfluorophenylmagnesium halides* 124
*Perfluoropyridine*
see perfluoroaromatics
*Perfluorovinyl organometallics* 123, 127
*Phenols*
reactions with fluoro-olefins 112, 133
*Phenylsulfur trifluoride*
fluorinating agent 41
preparation 15
*Phenyl trifluoroacetate* 118
*Phosphines, fluorinated*
nomenclature 19
preparation 129
*Phosphoranes, fluorinated*
fluorinating agents 40
nomenclature 19
preparation 25

*Phosphorochloridates, Phosphorofluoridates* see *Chlorophosphates, Fluorophosphates*
*Phosphorus halides*
  reaction with fluoroalcohols and acids 102
*Photoreactions* 146
*Physical constants and properties*
  of Freons 72
  of inorganic fluorides 16
  of monomers 73
  of organic fluorides 57–63
  of polymers 75
  of refrigerants 72
*Physicochemical properties* 63–67
*Physiological properties* 66–68
*Plastics*
  resistance to fluorine and hydrogen fluoride 6, 7
*Plastics, fluorinated* 70–77
*Platinum*
  catalyst for hydrogenation 89, 90
  resistance to fluorine and hydrogen fluoride 7
*Polar additions* to fluoro-olefins 105, 133, 134
*Polyethylene*
  comparison with fluoroplastics 75
  resistance to hydrogen fluoride 6, 7
*Polyhalides*
  conversion to fluorides 32, 34, 35
*Polymerization of fluorinated monomers* 74
*Polymers, fluorinated* 74–77
  pyrolysis 147
*Polytetrafluoroethylene*
  applications 6, 76, 77
  preparation and properties 74–76
    pyrolysis 147
  resistance to chemicals 7, 76
*Potassium fluoride*
  applications 16
  availability 9–12
  preparation 15
  properties 16
  replacement of
    halogens by fluorine 32, 36, 37
    sulfonyloxy group by fluorine 39
*Potassium fluorosulfinate*
  availability 9–12
  fluorinating agent 36
*Potassium permanganate*
  oxidation of fluoro compounds 96, 97
*Preparation*
  of inorganic fluorides 12–15
  of organic fluorides 21–45, 148–150
*Propellants* 69, 70
*Properties*
  of inorganic fluorides 14, 16
  of organic fluorides 57–68
  of polymers 75, 76
*Pseudoperfluoroalcohols* 77
*Pyrolysis of fluoro compounds* 147

**Q**

*Qualitative tests for fluorine* 50–52
*Quantitative determination of fluorides* 52–55
*Quarpel* 79

**R**

*Ratsbane*
  toxic principle 67
*Reaction heats of fluorination* 29
*Reactions of organic fluoro compounds* 80–157
*Reactivity of fluorine* 86, 106
  of halogens in Friedel-Crafts synthesis 106
*Reagents for fluoride detection* 51
*Rearrangements, molecular*
  of fluoro compounds 88, 123, 127, 145–148
*Reduction of fluoro compounds* 88–94
*Reformatsky synthesis* 128
*Refraction, atomic*
  of fluorine 60, 61
*Refractive index*
  of fluoro compounds 60, 61
  of fluoro polymers 75
*Refrigerants*
  applications 69–71
  nomenclature 19
*Replacement by fluorine*
  of amino group 42, 43
  of halogens 31–38
  of hydrogen 28–30
  of hydroxyl 39
  of oxygen 38–42

*Replacement by halogen*
of amino group 102
of fluorine 102, 103
of hydrogen 99, 100
of hydroxyl 102
*Resistance*
of materials toward fluorine and hydrogen fluoride 7
*Resistivity (specific)*
of fluoro polymers 75
*Rubidium fluoride*
applications 16
availability 9–12
preparation 15
properties 16
replacement of halogen by fluorine 37, 38

S

*Safety measures in fluorine chemistry* 5, 6
*Salts*—see the corresponding acids
*Sarin* 66, 78
*Schiemann reaction* 43
*Schöniger method* 54
*Scotchgard* 79
*Selenium tetrafluoride*
fluorinating agent 41
*Shifts of double bonds* 145
*Silicon–halogen bond*
cleavage by hydrofluoric acid 32, 33
*Silver*
catalyst for fluorinations 29
resistance toward fluorine and hydrogen fluoride 7
*Silver fluorides*
applications 16, 32
availability 9–12
fluorination 25, 26, 30
preparation 15
properties 16
replacement of halogen by fluorine 32, 35
*Silver permanganate*
combustion catalyst for analysis 52, 55
$S_N2$, $S_N2'$ *reactions* 91, 92
*Sodium*
decomposition of organic fluorides 51, 52
reduction of fluoro compounds 93, 94
*Sodium alkoxides*
catalysts for addition of alcohols to fluoro-olefins 133
reaction with perfluoroaromatics 116
*Sodium borohydride*
reduction of fluoro compounds 89
*Sodium dichromate*
oxidation of fluoro compounds 97
*Sodium fluoride*
applications 16, 37
availability 9–12
preparation 15
properties 16
*Sodium fluoroborate*
applications 16, 43
availability 9–12
preparation 15
properties 16
*Sodium fluorosilicate*
availability 9–12
replacement of chlorine by fluorine 37, 38
*Sodium hydride*
condensing agent in Claisen condensation 132
*Soman* 66, 78
*Spectrophotometric determination of fluoride* 54
*Stability of fluoro compounds* 63
*Stain-repellent properties* 79
*Steric effect of fluorine* 85
*Steroids, fluorinated*
preparation 31, 35, 41
properties and uses 78
*Substitution*
see *Replacement*
*Substitutions*
electrophilic 105, 106
nucleophilic 106–132
*Sulfonation of fluoroaromatics* 104
*Sulfones, Sulfoxides*
by oxidation of sulfides 97
*Sulfonyl chlorides*
conversion to fluorides 36
*Sulfonyl esters*
conversion to fluorides 39
*Sulfonyl fluorides*
conversion of alcohols to fluorides 39
preparation 36, 39

*Sulfur compounds*
electrochemical fluorination 25
*Sulfur dioxide*
compound with potassium fluoride 36
*Sulfur tetrafluoride*
application 16
availability 9–12
preparation 15
properties 16
reaction with cyanides 25, 26
replacement of oxygen by fluorine 40–42
*Sulfuric acid*
decomposition of organic fluorides 51, 52
*Surface tension of fluoro compounds* 61, 62
*Suspension polymerization* 74
*Synalar* 78

T

*Teflon* 1, 74, 76
*Telomers* 135
*Tensile strength of fluoroplastics* 75
*Tests for fluoride* 50–52
*Test tube for detection of fluoride* 51
*Thallium fluoride*
applications 16
availability 9–12
preparation 15
properties 16
replacement of halogen by fluorine 37, 38
*Thermodynamic properties of Freons* 72
*Thionyl chloride*
replacement of hydroxyl 102
*Thiophenols*
reactions with fluorohalo compounds 114
*Thorium alizarin lake*
detection of fluorine 51
*Thorium nitrate*
titration of fluoride 53, 54
*Tin*
reduction of fluoro compounds 93, 94
*Toluenesulfonates*
conversion to fluorides 39
*Tosyl group*
replacement by fluorine 39
*Toxicity of fluoro compounds* 67–69
*Trademarks*
of drugs 77, 78
of plastics 76
of refrigerants 71
*Tranquilizers, fluorinated* 78
*Transesterification of fluoroalcohols* 107
*Transition temperature of fluoro polymers* 75
*Triamcinolone* 78
*Trifluoroacetic acid*
see *Fluoroacids, Perfluoroacids*
*Trifluoroacetic anhydride*
acylation of
alcohols 118
amino acids 119
hydrogen sulfide 119
catalyst for esterification 106, 107, 152
reagent 152
*Trifluoroacetylation* 118, 119
*Trifluoromethyl group*
by conversion of carboxyl 41, 42
hydrolysis 110, 111
*Trifluoroperoxyacetic acid* 152, 153
*Trifluorothiolacetate, ethyl*
trifluoroacetylating agent 119
*Trifluorovinyl organometallics* 123, 127
*Triphenyldifluorophosphorane*
fluorinating agent 40
*Triphenylphosphine*
Wittig reaction 130
*Triple bond*
see *Acetylenes, Fluoroacetylenes*

V

*Vapor pressure*
of Freons 62, 72
of hydrogen fluoride 14, 17
*Velocity constants*
of displacement of fluorine by carboxylate 87
of displacement of iodine by thiophenoxide 81
*Vesprin* 78
*Viscosity of fluoro compounds* 63
*Viton A* 74
*Volumetric determination of fluoride* 52, 53

W

*Water*

*Water* (cont.)
addition to fluoro ketones 81
determination in
hydrogen fluoride 55
Freon 55
*Water repellents* 79
*Wickbold apparatus and method*
for determination of fluorine 54
*Wolff rearrangement* 146
*Wurtz-Fittig reaction* 129

**Z**

*Zepel* 79
*Zinc*
dehalogenation 142, 143
organozinc compounds 128
reduction of fluoro compounds 93, 94
*Zirconium alizarin lake*
test for fluoride 51